JOHN A. ROEBLING
AND HIS **SUSPENSION BRIDGE** ON THE OHIO RIVER

JOHN A. ROEBLING
AND HIS **SUSPENSION BRIDGE** ON THE OHIO RIVER

BY **DON HEINRICH TOLZMANN**

LITTLE MIAMI PUBLISHING CO·
Milford, Ohio
2007

Little Miami Publishing Co.
P.O. Box 588
Milford, Ohio 45150-0588
www.littlemiamibooks.com

Copies of this book can be obtained by contacting the publisher.

12 11 10 09 08   3 4 5 6

Photographs on the cover and pages ii, 48, and 49 provided by Stevie Publishing, Inc./Robert W. Stevie with his permission.

Printed in United States of America

ISBN-13: 978-1-932250-47-3

Library of Congress Control Number: 2007926746

CONTENTS

PREFACE

Its grace, beauty, and charm have adorned the Ohio River Valley for more than a century. Everyone in the area knows it's there, and visitors often come down to the riverbank just to admire it: the John A. Roebling Suspension Bridge on the Ohio River. Since its completion in the nineteenth century, it has become the major landmark of the region, as well as the veritable symbol of the Greater Cincinnati area. More than a landmark and symbol, however, it also served as the prototype of the Brooklyn Bridge, and was the largest of its kind when completed. Today, more than twenty thousand vehicles cross it daily, and it is the most popular pedestrian bridge on the Ohio River, but the Brooklyn Bridge is obviously more well known, rather than its model.

It is also not surprising that most works about Roebling tend to focus on the Brooklyn Bridge and its construction, and only briefly mention his other bridges, including the Suspension Bridge on the Ohio. Those that do discuss them concentrate mainly on bridge construction and related technical aspects. My goal is to illuminate not only the Roebling Suspension Bridge, but also the man who built it. The bridge builder belongs in the foreground, not in the background, since without the creator there would have been no creation.

This book emerged from an essay I published in 1998 in the series *Max Kade Occasional Papers in German American Studies*. Because of the considerable interest in Roebling, and because 2006 marked the 200th anniversary of Roebling's birth, as well as the 140th anniversary of the Suspension Bridge, I continued research on the topic. Hopefully, this work will contribute to an appreciation, not only of the Suspension Bridge, but its creator as well, and the important role they both play in the history of the region.

Many thanks to the following for providing information, illustrations, and photographs for this publication: John Banner, Manfred Schnetzer, Robert Stevie, Michael Weiss, the City of Mühlhausen, and the Little Miami Publishing Company.

Many thanks to my wife, Patricia, who read the final draft of this work. And, thanks to the Little Miami Publishing Co. for publishing this and other works of mine that aim to illuminate the German heritage of the Ohio and Mississippi River Valleys.

Don Heinrich Tolzmann

CHAPTER ONE

ORIGINS OF THE BRIDGE BUILDER

Beginnings

John A. Roebling, or Johann August, was born 12 June 1806 at Mühlhausen in Thuringia, the son of Christoph Polycarpus and Friederike Dorothea Roebling, with family ancestry reaching back to sixteenth century (Nicholaus Roebling, or Rebeling, born 1560). His education began at the Gymnasium in Mühlhausen, and continued at the Realschule in Erfurt. After these elementary studies, Roebling went to Berlin, where he studied engineering at the well-known Royal Polytechnical Institute (RPI). His family, which was involved in the tobacco industry, was not well-to-do, but, nonetheless, able to support his studies.

In addition to his studies in engineering, Roebling also attended the lectures of well-known philosopher Georg Wilhelm Hegel and it is often said that he was Hegel's favorite student.

It was Hegel's lectures on the history of philosophy that first brought America to Roebling's attention as the land of freedom and opportunity. Hegel stated that America was the land of the future, "where, in the ages that lie before us, the burden of the world's history shall reveal itself—perhaps in a contest between North and South America. It is a land of desire for all those who are weary of the historical lumber-room of old Europe." One admonition of Hegel's remained with Roebling forever: "One thing remember, nothing great in the world has been accomplished without passion."[1] There is no question that Hegel exerted a great influence on the young student.

After receiving a degree in civil engineering, Roebling took a position in 1827 with the Prussian government to work on road construction in Westphalia. His interest in bridges led him to complete a thesis focusing on the suspension bridge at Bamberg, It might be noted that a suspension bridge simply is a bridge that is suspended by cables that are hung over towers and then fastened on both ends. His bridge studies at the RPI and his thesis on the Bamberg suspension bridge influenced him greatly, as was later so clearly demonstrated in the United States. However, these were not the only influences on Roebling's life at the time.

After two years of road construction work, Roebling resigned his position to establish a bookstore and a book publishing company in Eschwege, an indication of his intellectual orientation. Here Roebling came to meet Johann A. Etzler, a Protestant minister, who was known for his interests in the natural sciences and America—two topics of great interest to the young Roebling. Etzler had spent the years 1822 to 1829 in the United States and then returned to Germany, only to be arrested on the charge of promoting political unrest by urging others to immigrate to America.

It was at this time, in the 1830s, that the great immigration waves from Germany were beginning, and many would come from the northwestern region where Roebling was situated. This was the period after the Napoleonic wars, when many sought a better economic and political situation in the New World, which was being praised as "the El Dorado" of the German immigrant.

Roebling and his brother, Carl Friedrich, read everything they could locate about America, including the highly influential report by Gottfried Duden, who had lived in Missouri and then written a book on the topic: *Bericht über eine Reise nach den westlichen Staaten Nordamerika's und einen mehrjährigen Aufenthalt am Missouri (in den Jahren 1824,*

25, 26 und 1827), in Bezug auf Auswanderung und Überbevölkerung, oder:
Das Leben im Innern der Vereinigsten Staaten und dessen Bedeutung für die
häusliche und politische Lage der Europäer . . . (Elberfeld: S. Lucas, 1829).
This travel report, simply referred to as the *Bericht*, or *Report*, was considered the most influential single book in the history of the German
immigration to America, and painted not only a positive image of the
New World, but a relatively glowing one as well.[2]

At this time, many groups and societies were being formed to or-
ganize the immigration to America, and so it was not surprising that
in Thuringia an emigration society was formed with its headquarters at
Mühlhausen. Friedrich Christoph von Dachroeden initiated it, and the
group naturally attracted quite a following. Among its leaders were Etzler, Roebling, Heinrich Harseim, and Emil Angelrodt (later a merchant
in St. Louis). The group soon had several hundred members, mainly
from the Mühlhausen and Darmstadt area.

Shortly after Etzler's release from prison in 1829, he collaborated
with Roebling and published a volume through the latter's press entitled
Allgemeine Ansicht der Vereinigten Staaten von Nord-Amerika für Auswan-
derer, nebst Plan zu einer gemeinschaftlichen Siedlung daselbst (Eschwege:
Gedruckt in der Röblingischen Buchdruckerei, 1830). This work aimed
to provide a general survey of the United States, and discussed the pos-
sibility of establishing a German colony in America, most likely in the
south (Florida, Louisiana, or Arkansas). In 1831, a second expanded
edition of the work was published. After actually encountering the re-
alities of life in the United States, Roebling would criticize this work,
stating that he wished it had never been published since it proved to be
so influential and persuasive in convincing others to immigrate.

He later wrote of the book, "the hardships that are connected with
emigration, especially for one who is taking the first step, are not given
enough prominence therein. I blame Etzler's carelessness and bold, un-
founded assertions for this." Roebling was even so concerned that he
proposed to write another, more realistic book, stating, "In the future I
will gather material for a true presentation of conditions in this country
and will include good advice for emigrants based on experience, and
I will make it my duty to have it printed for publication in German."

Moreover, he wrote that he would not, "for any price, persuade any person to come here, even if we should have to be here alone for the rest of our lives—we would not want to deceive anybody, as much as we desire to have our countrymen with us." Roebling, in short, felt that Etzler had provided an inaccurate account, and also that the south was not the place for Germans to settle for two reasons: the climate and the institution of slavery. Indeed, he wrote, "to take poor Germans there to work was a laughable idea of Etzler's—it would be absolutely impossible—none would stay there, not even for high wages." Both factors, the climate and slavery, he felt, would be repulsive to the German immigrant. Regardless of his later critique, the book, published through his press in Eschwege, would convince more people to join the Mühlhausen Emigration Society.[3]

The Roebling brothers, entrusted with six thousand dollars, sailed from Bremen on 21 May 1831, and landed 6 August in the port city of Philadelphia. After extensively investigating possible settlement sites, the Roeblings decided on a location in Butler County, Pennsylvania, in the western area of the state, just north of Pittsburgh. This was a much better site than those mentioned in Etzler's book, and the Roeblings had now the chance not only to examine sites, but to speak with other German-Americans on what the best locations could be.

Pennsylvania was an obvious location, as it had been the center of German settlement since the early colonial period, and the state was now one-third German-American. Moreover, there were quite a few German-American settlements in and around Pittsburgh. Across the border in Ohio one found settlements such as Steubenville. And in Pennsylvania there were a number of settlements that had been established by German immigration societies, such as the settlement at Economy, founded by Georg (*Vater*) Rapp.[4]

Rapp had led a group of Swabian immigrants from Iptingen in Württemberg to found a utopian religious community at Economy, eighteen miles north of Pittsburgh. Although highly successful economically, the community stood under the authoritarian rule of Rapp. In October 1831, Roebling visited Economy in search of a settlement site for his society. He noted that "everything is purposefully arranged and shows contentment and prosperity," but complained that Rapp preached "unreasonable stuff," such as celibacy, and correctly foresaw that such a utopian community was headed for a crisis situation. Later on it did experience a schism, resulting in the withdrawal of some of its members.

It was clear that Roebling wanted to establish a settlement in which each stood on his or her own independently, free of domination. He made it clear that what he sought in America was "a free, reasonable, natural relationship of the people toward each other; freedom and equality." America, in his view, was for those who were "for freedom and equal rights" and for those who would depend on their own individual strength and ability.

A Community Founder

Roebling and his colleagues purchased sixteen hundred acres of land in Butler County, about twenty miles north of Pittsburgh. In 1832, members of the Mühlhausen Emigration Society traveled to Bremen and set sail for Pennsylvania. It should be noted that Roebling had already demonstrated his inventiveness and originality in Germany, and was eager for the opportunity to develop his new ideas in America, where he would be free of red tape and bureaucracy. The Roeblings felt that agriculture would be the best field for them in the United States, and it was their plan to locate their settlement in the North. Here the climate was good, there was no slavery, and other Germans had settled in great number.

Their new settlement was first known as Germania, and then later as Saxonburg, and the settlers kept in touch with their friends and relatives in Germany, causing them to immigrate to Pennsylvania. In November and December 1831, Roebling wrote several extensive letters to friends in Germany—one of them more than one hundred pages in length. These letters described in detail the path Roebling had taken upon his arrival in the United States, and how one could get to Saxonburg.[5]

For example, he provided the exact information on the various port cities (New York, Philadelphia, Baltimore, and New Orleans), and how to get from them to the new settlement. Information was even provided about whom one should contact in Pittsburgh. He also explained in detail what to bring, and what would be needed here, and what the cost of various items would be in America.

The extensive detail and advice he provided to prospective immigrants in the homeland is impressive. He also wrote that Butler County was eager to attract German immigrants. He claimed, "if this region is built up by industrious Germans, then it can become an earthly paradise." He wrote in his letters that the little town would thrive and flourish, "like a plant whose growth was forced by diligent industry and farming," and that if one cleared land, one could use it for free for five years, and thereafter only pay one-third of the produce of the cleared land. He also advised: "Do not trust anyone whom you do not know very well—be careful—above all things, keep your money; and tell everyone to do the same."

He advised which trades would be needed, even writing, "Tell my cousin, the printer, that perhaps later on there will be an opportunity to establish a German printing and lithographic business here." Also, there would also soon be a need for a German-language school for the children. And Roebling stated, "Once you are here, you will not regret having left Germany." Many from his region heeded his calls, as stated in his 1831 book and extensive letters.[6] Among the arrivals was Ernst Herting, a tailor, who came in 1834. Two years later Roebling married his eldest daughter, Johanna, and their first son, Washington Augustus, was born at Saxonburg in 1837.

Roebling was a man bubbling with energy and ideas—by day he was a farmer tilling the soil, and by night he was an engineer coming up with all kinds of ideas for various inventions. He, quite understandingly, yearned to get back to his profession as an engineer that he had studied for in Berlin, and in 1837 obtained a position working on canal projects in Pennsylvania. At the same time, he also obtained U.S. citizenship.

Roebling maintained a lifelong interest in the German heritage, and raised his children bilingually. In 1838 and 1839, he attended the German-American Conventions at Pittsburgh as a delegate from Butler County. These were the first national conferences of German-Americans, and were attended by delegates from across the United States. As a result of the meetings, a German-language school was established at

Roebling's naturalization papers. He became a U.S. citizen in 1837, after settling in Butler County, Pennsylvania. [Hamilton Schuyler, *Roeblings, Bridge-Builders and Industrialists: The Story of Three Industrialists: The Story of Three Generations of an Illustrious Family, 1831–1931* (1931).]

Phillipsburg, Pennsylvania, and other plans were devised for the preservation of the German heritage.

Roebling became interested in the Allegheny Railroad, an important link between eastern and western Pennsylvania. He noted that the canal boats had to be moved by cables up and down the steeply inclined railway, and that the cables were constructed of Kentucky hemp, generally about three inches in diameter. This concerned him after he witnessed an accident caused by the frayed hemp rope on the railroad, and he remembered reading about an inventor in Saxony who had twisted wires together to form a rope.

Beginnings of the Bridge Builder

By means of his studies, Roebling had noted the weakness of bridges that made use of ropes or chains and came up with the brilliant idea of making use of cables consisting of numerous wires braided together. In 1841, he began making wire at his factory in Saxonburg, and in 1843 he published an article in the *American Railroad Journal* about the value and uses of his new product. In 1844, he then built a wooden aqueduct for the Pennsylvania Canal that made use of wire cables and completed a suspension bridge on the Monongehala River at Pittsburgh in 1846. In 1848, Roebling moved his wire factory to Trenton, New Jersey. Altogether, by 1850, Roebling had built a total of four suspension bridges.

The original Trenton, New Jersey, factory of the Roeblings in 1849. [Illustration from Hamilton Schuyler, *Roeblings, Bridge-Builders and Industrialists: The Story of Three Industrialists: The Story of Three Generations of an Illustrious Family, 1831–1931 (1931).*]

The idea of a bridge on the Ohio River had been discussed for some time in the early nineteenth century. The desire for the bridge originally came from the citizens of Lexington, Kentucky, a town that saw its economy falling behind that of various Ohio River cities when the steamboats began their commerce on the river. Lexington's citizens felt that there needed to be a north–south connection between Ohio and Kentucky. They met in 1839 to discuss this need, and the plan was diverted into a turnpike from Covington, Kentucky. However, the creation of the turnpike did not quell the interest in a bridge. In February of 1846, a group of citizens in Covington requested and obtained a charter

Röblingstrasse in Mühlhausen. A street named in honor of Roebling by his hometown. [Courtesy of Michael Weiss.]

Poster to an exhibit held in Mühlhausen in honor of the 200th anniversary of Roebling's birth. [Courtesy of Michael Weiss.]

Roebling's birthplace in Mühlhausen.

Ancestral home of the Roebling family in Mühlhausen. [Courtesy of Michael Weiss.]

from the Kentucky General Assembly that incorporated the Covington and Cincinnati Bridge Company.

In May 1846, the bridge committee invited Roebling to survey the area, and, subsequently, he published a brilliant report analyzing the technical and commercial aspects and possibilities relating to a bridge on the river. The bridge company accepted his report, but the company also had to receive a charter from the State of Ohio, and north of the Ohio River there was plenty of opposition to the plan.

Opponents of the proposed bridge got together and published an anonymous pamphlet denouncing the idea of a bridge; it was, therefore, not surprising that the Ohio State Legislature in 1847 failed to approve the charter for the bridge company. However, in 1849 it changed its tune, when work on the suspension bridge at Wheeling, West Virginia, neared completion. The State of Ohio approved the charter that Kentucky had endorsed two years earlier, but attached a rider so that no street in Cincinnati could be built in line with the proposed bridge. This stipulation satisfied commercial interests in Cincinnati, who feared that they would lose business to northern Kentucky. It should be noted that a major supporter of the bridge plan in the Ohio Legislature was Carl Reemelin, a German-American state senator from Hamilton County, who had met with Roebling in Cincinnati and become a strong supporter of his plans for the bridge.[7]

Businesses and companies afraid of losing business to Kentucky were not alone in opposing the bridge. Indeed, there were political, as well as commercial, arguments against the idea of a bridge. There were some in the Ohio Legislature who felt that such a bridge would lead to a flight of slaves from the South to the North, as the Underground Railroad was, by now, well underway. Steamboat companies also opposed a bridge, fearing that it would hurt their business. And others claimed that such a bridge would cause ice jams and floods. In short, there were a variety of forces that stood in opposition to the idea of building a bridge on the Ohio River.

An additional factor working against suspension bridges was the public's lack of trust in the strength of suspension bridges. Also, railroads were now coming on the scene, and investors were looking more

Bronze historical marker at Roebling's birthplace in Mühlhausen. [Illustration courtesy of the City of Mühlhausen, Germany.]

at them than bridges as places in which to invest.

In the meantime, Roebling was engaged with his various business projects and publications. He saw that railways would soon surpass the canal systems as the most effective mode of transportation and, in 1847, read a paper before the Pittsburgh Board of Trade in which he advocated the construction of a railway through the state of Pennsylvania. In 1850, Roebling wrote in the *Journal of Commerce* that plans should be undertaken to construct a transatlantic telegraph, and estimated that the cost would be $1,300,000. These two articles attracted attention, and indicated the kind of insight that Roebling displayed in the field of engineering.

A musical arrangement in the form of a march was composed to commorate the opening of the Suspension Bridge in 1866. [Library of Congress]

CHAPTER TWO

THE **SUSPENSION** BRIDGE

The Covington and Cincinnati Bridge Company

John Roebling was now acquiring a national reputation as a bridge builder. He pioneered railroad suspension bridges with the construction of the bridge at Niagara Falls from 1851 to 1855. Most American engineers had considered the plan for a railway suspension bridge predestined to failure, since no such bridges had ever been built. In 1856, Roebling's plans for a bridge across the Ohio River connecting Cincinnati and Covington were accepted. He signed the contract and the work was begun.[8]

The company was led by a businessman of German descent, Amos Shinkle (1818–1892). His entrepreneurial accomplishments included the establishment of a bank, the construction of steamboats, the introduction of gas lighting and telephone service in Covington, and the building of more than thirty houses in Covington. In 1856, Shinkle became a member of the Covington and Cincinnati Bridge Company, and in 1866 became president, an office he held until his death in 1892.

The work came to a standstill with the advent of the Civil War, and the bridge could not be completed until 1867. Work, however, was be-

One of Roebling's plans for the Suspension Bridge on the Ohio River—this one with Gothic towers.
[Illustration courtesy of the Rensselaer Institute Archives and Special Collections.]

gun in August 1856, as the stone, cement and equipment had been purchased. Roebling already had selected Buena Vista sandstone from two
hundred miles away, since it contained petroleum that gave it a greater
water resistance. By 1857, 166,000 cubic feet of stone had been laid.
Limestone for the towers came from nearby Dayton, Ohio. Work then
halted due to the weather, as well as the 1857 financial panic.

It was during this time before the Civil War that Heinrich A. Rattermann, a prominent German-American historian in Cincinnati, came

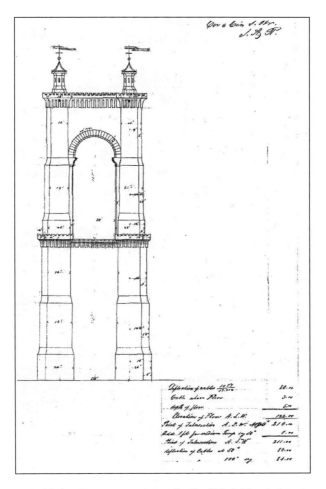

Roebling eventually decided on Romanesque towers for the Ohio bridge, but made use of the Gothic style with plans for the Brooklyn Bridge. [Illustration courtesy of the Rensselaer Institute Archives and Special Collections.]

to know Roebling, and later wrote an article about him. Roebling enjoyed social get-togethers in the German Over-the-Rhine district, according to Rattermann. Here he often discussed the bridge and all of its technicalities in detail. This had the eventual result that "the Germans of Cincinnati soon were added to the strongest supporters not only of the undertaking, but also of Roebling's ideas."[9]

They demonstrated this by investing in the bridge company as a show of support.

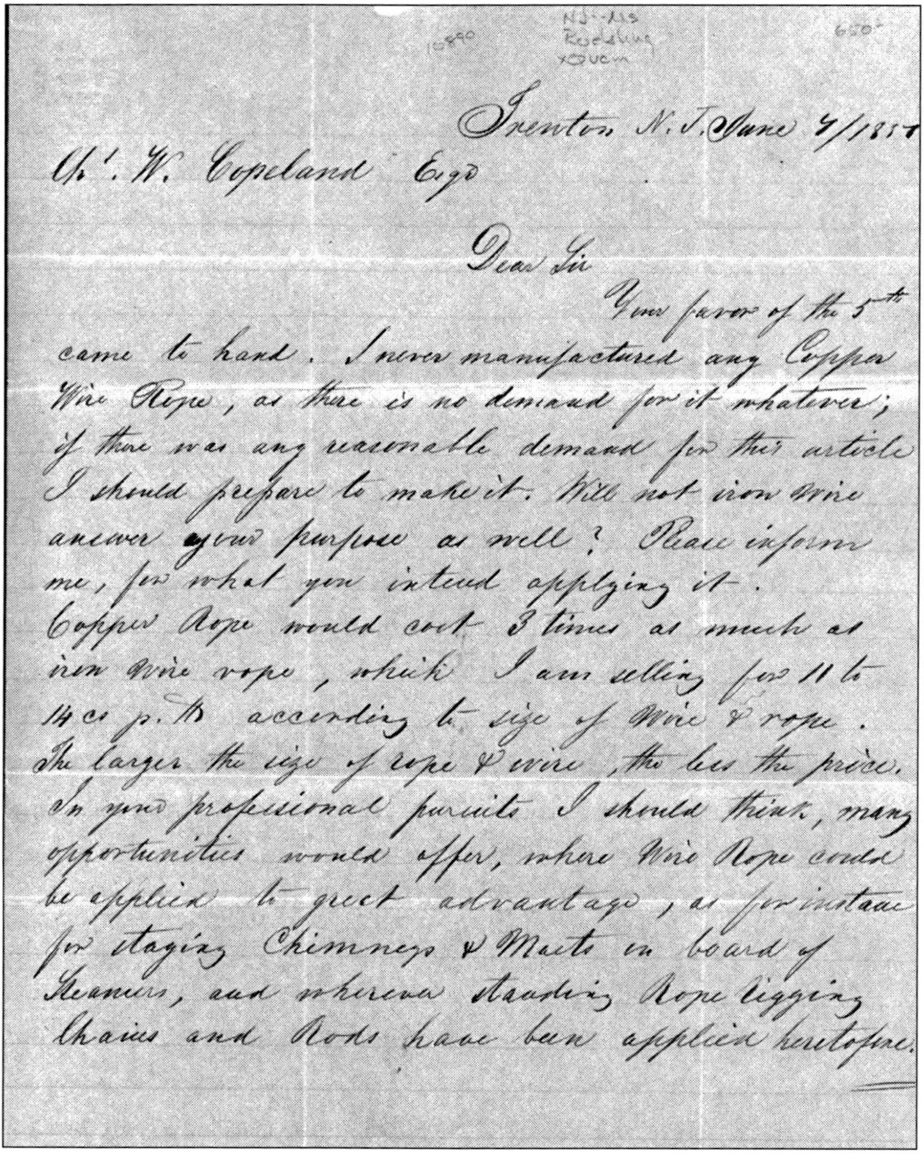

Trenton N.J. June 7/1851

Mr. W. Copeland Esqr

Dear Sir

Your favor of the 5th
came to hand. I never manufactured any Copper
Wire Rope, as there is no demand for it whatever;
if there was any reasonable demand for this article
I should prepare to make it. Will not iron wire
answer your purpose as well? Please inform
me, for what you intend applying it.
Copper Rope would cost 3 times as much as
iron wire rope, which I am selling for 11 to
14 cts p. lb according to size of wire & rope.
The larger the size of rope & wire, the less the price.
In your professional pursuits I should think, many
opportunities would offer, where Wire Rope could
be applied to great advantage, as for instance
for staying Chimneys & Masts on board of
Steamers, and wherever standing Rope rigging
Chains and Rods have been applied heretofore.

In 2001, the German-Americana Collection at the University of Cincinnati obtained an 1857 letter written by Roebling which discusses the benefits of iron wire in relation to copper wire. [Illustration courtesy of the German-Americana Collection, University of Cincinnati.]

Page from John Roebling's workbook. [Courtesy of Rensselaer Institute Archives and Special Collections.]

Illustration depicting the volunteers crossing the Ohio River from Cincinnati, Ohio, to Covington, Kentucky, on the pontoon bridge. [Illustration provided by the Little Miami Publishing Co.]

Rattermann recalled get-togethers with Roebling and Peter Kaufmann at the *Weinstube* of Nicholaus Schmitt in the Over-the-Rhine district. Kaufmann had belonged to the utopian settlement of Rapp's at Economy, Pennsylvania, before he founded his own settlement in Teutonia, Ohio. Later, he moved to Canton, where he published a German-American newspaper. Kaufmann, like Roebling, was deeply interested in German philosophy, especially the work of Hegel, and published a well-known book in 1858, which was imbued with the Hegelian spirit. It appeared in German as *Der Tempel der Wahrheit* . . . (Cincinnati: Truman & Spofford, 1858), as well as in English as *The Temple of Truth* . . . (Cincinnati: Truman & Spofford, 1858). According to Rattermann, the three friends often met in the Over-the-Rhine district, where they discussed philosophy and other topics at length.[10]

The Civil War and Thereafter

Although the outbreak of the Civil War brought work on the bridge

Wire plant of John A. Roebling's Sons Co. in Trenton, New Jersey, c. 1906 [Photograph courtesy of Little Miami Publishing Co.]

A popular nineteenth-century lithograph of the Suspension Bridge. [Author's Collection.]

to a standstill, it also brought home the need to complete it, especially after the July 1862 threat caused by the arrival of John Hunt Morgan's raiders. Area volunteers made it across the river by means of a pontoon bridge and chased the Confederates off, but the event raised area determination to get the job done, and also caused the Ohio Legislature to pass an amendment in support thereof. However, workers and supplies were hard to come by during the war, but by September 1865, wires were being run across the river, making it possible to produce the well-known massive cables measuring twelve and a half inches. Suspenders were attached, a wooden bed of oak and pine laid into place, and, by the following year, the work was finally done!

The Suspension Bridge was opened to pedestrians on 1 December 1866, and in two days, 120,000 people had paid three cents to cross over it. Roebling and Amos Shinkle, president of the Bridge Company and the major proponent of the bridge, led the official parade on New Year's Day. It had taken ten years to build the bridge and taken more than twice the funds that Roebling had projected, but everyone was pleased with it. According to David McCullough "it was unquestionably the finest as well as the largest bridge of its kind built until that time. Both structurally and architecturally it was a triumph."[11]

By now, Washington Augustus Roebling (1837–1926) was playing a major role in assisting his father. In the Civil War, he had been promoted to the rank of lieutenant colonel for gallantry and meritorious service. Not surprisingly, he had served as an engineer charged with the task of bridge building, and thereafter, helped his father complete the bridge on the Ohio River. After studies in Europe, he then returned to work with his father on the Brooklyn Bridge.[12]

Also, it should be noted that Washington was the builder and chief engineer for the Brooklyn Bridge after his father died.

Roebling's wire factory, established in Saxonburg in 1841, became the foundation for an industrial empire, The John A. Roebling's Sons Co. of Trenton, New Jersey, and was inherited by sons, grandsons, great-grandsons, etc. Roebling and his wife had nine children; seven of them were still at home at the time of his untimely death in 1869.

Another view of the Cincinnati Suspension Bridge. [Illustration courtesy Little Miami Publishing Co.]

Steamboats along the Cincinnati riverfront with the Suspension Bridge in the background pre-1897. [Author's Collection.]

A Source of Pride

Shortly after Roebling's death in 1869, *Der Deutsche Pionier*, the well-known German-American historical journal of Cincinnati, eulogized the great bridge builder: "If ever a pioneer in our adopted Fatherland has brought honor to the German name, then it certainly was John A. Roebling . . . who just recently passed away in the midst of his splendid work." State Senator Carl Reemelin of Cincinnati stated at the same time, "Thanks be to the Old Country for having sent us its son." Such eulogies typified the fact that Roebling, who was held in high regard across the United States by the American public because of his engineering genius, occupied a special place in the hearts and minds of German-Americans.[13] They were especially proud of Roebling for several reasons.

First, there was great pride in Roebling's engineering accomplishments, as exemplified by the suspension bridges he had constructed in the United States, and which were widely recognized as masterly achievements and historic landmarks.

Second, he represented what an immigrant could achieve in America, "the land of unlimited possibilities." Here he served as a role model and as an ideal to which others could aspire.

Third, the Roebling Suspension Bridge in Cincinnati symbolized German-American contributions to the building of the nation. German-Americans were actively involved in almost every field of human endeavor, most recently in the struggle on behalf of the Union in the Civil War.

Finally, the bridge itself served as a symbol for those coming down the river that the Greater Cincinnati area was, and had been since the early nineteenth century, a destination for immigrants as well as settlers. They knew when they caught sight of the bridge that they had arrived at their destination. Roebling and the bridge are, hence, inextricably connected with the history of the Greater Cincinnati area.

An Engineer's Report

The best place to find Roebling's own description of the bridge on the

Ohio River is in his "Report of John A. Roebling, Civil Engineer, to the President and Board of Directors of the Covington and Cincinnati Bridge Company, April 1st, 1867," which was published in the *Annual Report of the President and Directors to the Stockholders of the Covington & Cincinnati Bridge Company: For the Year Ending Feb. 28, 1867* (Trenton, New Jersey: Murphy & Bechtel, 1867).

From this lengthy report, the following comments by Roebling provide insight not only into the construction of the bridge, but also into Roebling's style of reporting:

At the lowest stage of water the Ohio River, between Cincinnati and Covington, has a width of about 1000 feet. By the charter of the company the position of the towers was fixed at the low-water mark, so that the middle span should present an opening of not less than 1000 feet in the clear. In the spring of the year 1832 the river rose sixty-two feet above low water. At this stage the width of waterway is over 2000 feet. With the exception of the towers, the whole waterway between the two cities is left unobstructed, a width of 1619 feet. The two small spans left open between the abutments and towers are each 281 feet, from face to center of towers. From an engineering point of view, this division of spans is not the most economical. The cheapest arrangement would have been one center span of 800 feet, and two half spans of 400 feet each. But that plan had been forestalled by previous legislation. One of the early charters decreed one single span of 1400 feet in the clear. But this very great and expensive span was afterwards allowed to be reduced to 1000 feet, and with this amendment the foundations were commenced in 1856. . . .

Owing to the persistent opposition of property owners, steamboat and ferry interests, the clear elevation of the floor above low-water mark, in the center of the river span had been fixed at 122 feet. By a later enactment, this height was reduced to 100 feet. As the bridge stands now, its elevation is 103 feet in the clear above low-water mark, at a medium temperature of sixty degrees, rising one foot by extreme cold and sinking one foot below this mark in the extreme heat. The greatest ascent is only five feet in one hundred, at the Cincinnati approach, and this diminishes as the suspended floor is reached . . . .

The floor of the bridge is formed of a strong wrought-iron frame, overlaid with several thicknesses of plank, and suspended to the two wire cables by means of suspenders attached every five feet, arranged between roadway and footpaths; the latter seven feet wide, and are protected by iron railings towards the river. The roadway is twenty feet wide, forming two tracks of four lines of iron trams, on which the wheels run, each tram being fourteen inches wide, to accommodate all kinds of gauges. The whole width of the floor between the outside railings is thirty-six feet. No stays or other obstructions are put up below the floor, such as may be seen under the Niagara bridge. No such means to prevent the floor from rising was used in this work; its security and stability are provided for by other appliances. The rock underneath the Niagara bridge afforded a very cheap mode of anchoring; it would have been a great oversight on my part not to avail myself of the under-floor stays in such a favorable locality. But in the Ohio River no such appendages were admissible. . . .

The general plan which I have always pursued in my works insures, by the heavy contraction of the cables in the center of the span, great lateral stability at this point. The larger and heavier the span, the greater will be its comparative stability at the center. Vertical stability in the center is also insured in large spans by the weight of the structure. But not so between the center and the towers. In consequence of the equilibrating tendency of the two opposite halves, vertical oscillations occur easier, and the great length of suspenders, acting like pendulums, promotes lateral displacements. These tendencies have to be met, and are thoroughly overcome in the Ohio bridge by an effective system of stays. The very careless manner in which stays have been attempted heretofore is a violation of the principle involved. Their arrangement in this bridge not only insures their own freedom from oscillation, but renders them fully effective by the uninterrupted preservation of their lines. . . .

Aside from simply stiffening the floor, the stays are rendering another and very important service; they effectually insure equilibrium between the main and half spans. Without stays the balance between adjoining spans would sometimes be greatly disturbed by unequal loads. The large crowds of many thousand of people, which frequently cover the floor from one end to the other, are occasionally very unevenly distributed, but they have never produced the slightest injurious effect upon the static condition of this work. . . .

Great doubts are yet entertained by many engineers, particularly in Europe, in regard to the fitness and safety of suspension bridges for railway purposes. By an additional expenditure of fifty thousand and a railroad track laid down in the center of the floor, the Ohio Bridge could have been made serviceable for the passage of locomotives and trains at the highest speed. Let any person who doubts this, observe the very slight tremor which is produced in this bridge by a long line of heavily loaded teams ten in a row, and he will readily understand that but a small addition of rigidity is wanted in order to pass railroad trains.

Bronze Statue of Roebling at Trenton, New Jersey.

CHAPTER THREE

ROEBLING

The Roebling Mystique

Roebling was a man of impressive stature, and it was said that his mere presence commanded respect. He was described as kind, modest, and friendly, but "his exterior did not betray such lovable characteristics." Roebling's "aquiline nose and firmly set mouth bore witness to a man with a strong will and a bold and enterprising spirit," and genius was said to flash from his deeply set eyes like lightning.[14] His whole countenance and demeanor indicated that he was not only a thoughtful man, but also a man of action, who was results-oriented.

Der Deutsche Pionier characterized him as a man of unshakable will power, penetrating originality, and as one for whom the word "failure" did not exist. Roebling carefully observed his time, and was known for his exacting punctuality. If a visitor was more than five minutes late, the appointment was canceled. Washington said of him, "The leading feature of my father's character was his intense activity and self-reliance. I cannot recall the moment when I saw him idle. . . ."[15]

Roebling was also a philanthropist who contributed generously to various beneficial causes. He greatly enjoyed music and took pleasure in playing the flute and piano. As a former student of Hegel, Roebling was greatly interested in philosophy, and planned on publishing a work dealing with philosophical matters. Among his manuscripts there is a work entitled: "A Metaphysical Essay on the Nature of Matter and of Spirit." Here he writes:

There is but One—all-powerful, all present, all wise and intelligent, pervading the whole universe. This One is independent of space and time, therefore has existed and will exist forever. . . . There is no chance. All is law and premeditated design, all is system and order. . . . There is one great book of Revelation—the book of Nature. . . . Our reasoning faculties are inspired by the Great Spirit of the Universe. . . . The arrangement of the whole universe is evidently calculated for progressive although slow improvement. As far as our mental and moral faculties are in our own power, we are free agents and responsible for our thoughts and action. Every sentiment of our mind forms a note of music of creation and will, either as dissonance or consonance, depress or swell the harmony of the whole.[16]

If this provides some insight into Roebling's *Weltanschauung*, then D.B. Steinman supplies some additional observations of note:

There was something striking in John Roebling's appearance, which the uninitiated might at first interpret as unfriendliness, but which in reality was only the outward aspect of a constant and intense inner concentration. Under the high forehead and deep under the shaggy brows lay his overshadowed gray eyes, eyes that rested on people and on things, firm and clear, earnest and penetrating. The strongly carved features, the firmly set lips, the rather haggard face with the cheeks somewhat sunken, the erect carriage of the tall man—all these betokened a strong personality and an inflexible power of determination.[17]

Roebling was known to be sparing with conversation, as he was often engrossed with his various plans and projects. Nonetheless, he was usually the first to exchange a hearty word of greetings. And, when engaged in discussion, he "spoke frankly and cheerfully, with a fresh and wholesome humor, giving himself naturally and without reserve."[18]

Many comments about Roebling relate to his brilliance, creativity, and extraordinary output. Above all, he was methodical, which was "a quality that reached into every corner of his life and profession. The collective result of these qualities was an outpouring of verbal and graphic material that is nothing less than astonishing, in terms even of what has survived. If we attempt to extrapolate from this to imagine the volume of material that does not survive—of which there must have been a considerable amount—the total product seemingly is more than that of five men's lives than one."[19]

Hamilton Schuyler's book on the Roebling family, *The Roeblings: A Century of Engineers, Bridge-builders and Industrialists: The Story of Three Generations of an Illustrious Family, 1831–1931*, provides further insight into the character and personality of Roebling. He notes that Roebling's energy knew no bounds and that his zeal never flagged. He also notes what a reading of Roebling correspondence reveals in the following:

There are few quotable passages in these hundreds of letters, little or nothing that would be of interest to the general reader as disclosing the personal and human side of the writer. Yet, apart from the light, which these letters throw upon the operations of the Roebling industry in the early days, they are still valuable as an indirect revelation of character. They show a man so intensely absorbed with his immediate tasks that he has neither the time nor the inclination for anything else. John A. Roebling as an intelligent and observant person must have had many unique experiences during his travels from place to place. He must have met many interesting people, but if so, no inkling of these things crop out in his letters. An individualist, self-contained and self-centered, he seems to have had no interests apart from his work. His energy was stupendous, his concentration perfect, his memory unfailing, his decisions prompt and irrevocable.[20]

In his work entitled *Lives and Works of Civil and Military Engineers*, Charles B. Stuart wrote of Roebling that his

reasoning was always clear, simple, and explicit, and sustained by philosoph-

ical and scientific facts. He took nothing for granted. His arguments were drawn from his store of scientific knowledge, with a mathematical accuracy and fitness that carried with them a conviction of truth. He was impatient at captious opposition to his projects, but always courted a discussion of his plans by those who brought sound theoretical or practical opposition to his views.[21]

Stuart also noted that:

One of his strongest moral traits was his power of will, not a will that was stubborn, but a certain spirit, tenacity of purpose, and confident reliance upon self, that was free of conceit; an instinctive faith in the resources of his art that no force of circumstance could divert him from carrying into effect a project once matured in his mind. His skill as an engineer was not surpassed by his exact probity. He held it 'to be the duty of an engineer, when charged with the design of public works, to report previous to their execution fairly, accurately, and candidly,' and that 'honesty of design and execution, next to knowledge and experience, most surely guarantees professional reputation.' Before entering upon any important work, he always demonstrated to the most minute detail its practicability, to his own mind at least, by scientific experiment and critical text; and when his own judgment was assured, no opposition, sarcasm, or pretended experience, could divert him from consummating his designs, and in his own way.[22]

CHAPTER FOUR

ROEBLING'S **LEGACY**

The Brooklyn Bridge

Whe Roebling Suspension Bridge served as the prototype for Roebling's next project, the Brooklyn Bridge. Plans for the bridge were first proposed in 1857, when Roebling suggested in a letter that a suspension bridge be constructed over the East River between lower Manhattan and Brooklyn. In 1869, Roebling perfected the plans and commenced work, but then suffered a tragic accident on 28 June that cost him his life.

While personally engaged in laying out the towers of the bridge, Roebling was injured by a falling piece of timber so that several of his toes had to be amputated. The operation was successful, but a few days later tetanus set in, to which Roebling finally succumbed at the age of sixty-three on 22 July. The task of completing the Brooklyn Bridge then fell to his son, Washington. It was dedicated on 24 May 1883, an event attended by the President of the United States, Chester A. Arthur.

Many of the comments made about the Brooklyn Bridge were similar to those made of the Roebling Suspension Bridge on the Ohio River. One commentator said, "The Bridge is beautiful in itself." It was noted that the Brooklyn Bridge "was more than an engineering feat and more than its physical beauty"; it had become "a uniting force. . . . It represented in a profound and visual way all that was best in America in 1883." And to the immigrants arriving at nearby Castle Gardens, "it was an immediate, visual proof of what an immigrant could achieve." It was well known that an immigrant had built the bridge, and that most of the labor force had been of immigrant stock. At the time of his death, an obituary in *Harper's Weekly* noted that Roebling "won the love and respect of all who knew him, and in his works has left behind him a nobler monument than could be shaped in marble."[23]

The Brooklyn Bridge has served as an inspiration to many. Engineers and writers, in particular, have marveled at its complexity and beauty. Ohio's greatest poet, Hart Crane, wrote of Roebling: "The man was a genius—and his accomplishments stupendous." In the 1920s, Crane lived in the same house in Brooklyn, and later on in the very same room from which Washington Roebling directed the completion of the Brooklyn Bridge. From Roebling's former dwelling, Crane wrote what is considered his greatest poem, *The Bridge*, published in 1930, in which he described the Bridge as a "harp and altar." While working on the poem, he wrote: "I'm working on a synthesis of America and its structural identity now, called *The Bridge*." A reviewer later noted that Crane treated the Bridge not only as a symbol, but also as a visible and tangible token of America and one that serves as a bridge to our past connecting us to the present and future.[24] The concluding stanzas that follow are from "To Brooklyn Bridge," which is one section of the longer work on the Bridge:

> O harp and altar, of the fury fused,
> (How could mere toil align thy choiring strings!)
> Terrific threshold of the prophet's pledge,
> Prayer of pariah, and the lover's cry,—
>
> Again, the traffic lights that skim thy swift
> Unfractioned idiom, immaculate sigh of stars,
> Beading thy path—condense eternity:
> And we have seen night lifted in thine arms.
>
> Under thy shadow by the piers I waited;

Washington Roebling.

President Chester A. Arthur and his party, crossing the Brooklyn Bridge at its opening in 1883. [Hamilton Schuyler, *Roeblings, Bridge-Builders and Industrialists: The Story of Three Industrialists: The Story of Three Generations of an Illustrious Family, 1831–1931* (1931).]

The Brooklyn Bridge. [Photograph provided by the Little Miami Publishing Co.]

Only in darkness is thy shadow clear.
The City's fiery parcels all undone,
Already snow submerges an iron year—

O sleepless as the river under thee,
Vaulting the sea, the prairie's dreaming sod,
Unto us lowliest sometime sweep, descend
And of the curveship lend a myth to God.

The Golden Gate Bridge

One historian has suggested, "that the history of engineering in the United States is almost identical with the history of the German-American engineers."[25] Many, of course, had been influenced by Roebling and the bridges that he built.

One was a young German-American student in Cincinnati by the name of Joseph Baermann Strauss (1870–1938). Born into a Bavarian immigrant family in Cincinnati, his father was an artist and his mother a musician. Although he maintained a lifelong interest in literature and poetry, his major interests were for the sciences and mathematics. In 1892, he obtained a degree in engineering from the University of Cincinnati. When he failed to make the football team due to his small stature, Strauss resolved that he would make his name in history for building something monumental and that he would become a bridgebuilder.

His fascination with bridges developed from his observation of the Roebling Suspension Bridge in Cincinnati. This bridge, no doubt, served as the inspiration for his decision to become a bridgebuilder, and to create something as monumental as Roebling's creation. Altogether Strauss built more than four hundred bridges in the United States, Cuba, the Canal Zone, Norway, Russia, and Japan. However, he is most well known as the chief engineer of the Golden Gate Bridge, which is considered his crowning achievement, and which was the longest single-span suspension bridge in the world when it was constructed.

When the Golden Gate Bridge was dedicated on 26 May 1937, two fellow Cincinnatians were also in attendance. Raymond Walters, president of the University of Cincinnati, and Alfred K. Nippert watched as

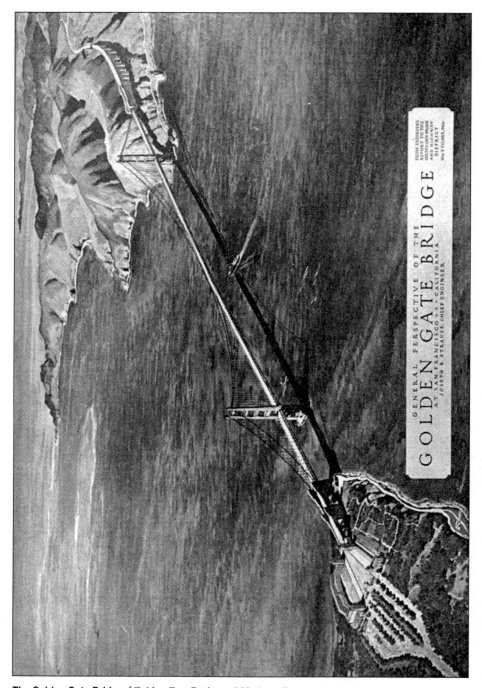

The Golden Gate Bridge. [Golden Gate Bridge and Highway District, *Golden Gate Bridge at San Francisco, California: Report of the Chief Engineer, with Architectural Studies, to the Board of Directors, Golden Gate Bridge and Highway District, August 27, 1930* (San Francisco, Calif.: The District, 1930)]

the U.S. Fleet, consisting of some forty battleships, cruisers, and some smaller craft, sailed under the bridge.

Strauss, like Roebling, was a man on a mission with vision, dedication, and conviction. His son, Richard Strauss, said that his father "defined success as a lot of hard work and not being discouraged by what other people said, but it was mostly hard work. He was always thinking. He stayed inside himself a lot of the time, just thinking. He was never without notepad and pencil—even kept it by his bedside because, as he said, 'You have to be ready for a good idea.'" His son also described his father as an incredible man, who was "very complex, indomitable, and tenacious, and had a profound desire to achieve.He was both a considerate and an impatient man. I never saw him really angry."[26] Like Roebling, his religious views were philosophical, rather than orthodox.

Rather than expressing his views by writing philosophical treatises, Strauss often expressed his views in poetry. For example, he composed a poem to commemorate the opening of the Golden Gate Bridge, two stanzas of which follow:

The Mighty Task is Done

At last, the mighty task is done;
Resplendent in the western sun,
The Bridge looms mountain high,
Its Titan piers grip ocean floor,
Its great steel arms link shore with shore,
Its towers pierce the sky.

On its broad decks, in rightful pride,
The world in swift parade shall ride,
Throughout all time to be,
Beneath, fleet ships from every port,
Vast landlocked bays, historic fort,
And drafting all the sea.[27]

The Golden Gate Bridge became known as one of the Seven Wonders of the World, and became one of the major U.S. landmarks, like the Brooklyn Bridge created by the Roebling family. Moreover, to spin the main suspension cables of the bridge, Strauss had hired none other

Tower perspective of the Golden Gate Bridge. [Golden Gate Bridge and Highway District, *Golden Gate Bridge at San Francisco, California: Report of the Chief Engineer, with Architectural Studies, to the Board of Directors, Golden Gate Bridge and Highway District, August 27, 1930* (San Francisco, Calif.: The District, 1930)]

than the company of Roebling & Sons, who supplied eighty thousand miles of wire, so that Roebling was not only involved in an inspirational capacity with Strauss; his family's company actually was involved in its construction as well.

On the occasion of the fiftieth anniversary of the Golden Gate Bridge, Joseph A. Steger, president of the University of Cincinnati, noted "it is with great pride that the University of Cincinnati honors Joseph B. Strauss during this Jubilee Celebration of the Golden Gate Bridge. He was a brilliant and inventive engineer who shared his genius and his determination with the world for nearly half a century . . . Strauss's capacity to dream, to dare what others would not, and to believe in his own ability to succeed was instilled and nurtured by the University of Cincinnati. He was an innovative and gifted student, and he was expected to make significant contributions." Steger also noted, "every age produces its own greatness. Each person appears at his or her appointed time to make a special contribution to the advancement of humankind. There is no predetermined line of succession—the scepter passes randomly."[28] In the case of Strauss, there is no doubt that it was the lasting legacy of Roebling that provided the necessary inspiration.

The Ohio River Valley Legacy

The Suspension Bridge on the Ohio River had cost $1.8 million to construct, and measured 1,619 feet from shore anchor to shore anchor. In 1867, pedestrians paid $1.25 for one hundred tickets, or three cents per crossing. In 1895, horse cars became obsolete as electric streetcars replaced them. In the same year, work began on refurbishing the bridge under the direction of Wilhelm Hildebrand, who had served as principal assistant engineer for the construction of the Brooklyn Bridge. The bridge then was painted blue and two steel cables were added to the iron ones already present, which doubled the strength of the bridge and prepared it well for the heavier use of the twentieth century.[29] The Covington–Cincinnati Bridge Company ran the bridge until it was purchased by the State of Kentucky in 1953 for $4.2 million. The dependability of the bridge was highlighted during the 1937 flood, when Roebling's Sus-

Toll-Gate, Cincinnati. [Illustration from the collection of Barbara Gargiulo, Little Miami Publishing Co.]

American Society of Civil Engineers
345 East 47th Street, New York, N.Y. 10017-2398

Brooklyn Bridge
1883 1983
USA 20c

Brooklyn Bridge, New York City
Designated by the
American Society of Civil Engineers
A National Historic
Civil Engineering Landmark

BROOKLYN, NY
MAY
17
1983
11201

FIRST DAY OF ISSUE

Planned and designed by
John A. Roebling (1806-1869)
A pioneer of civil engineering.

Washington A. Roebling (1837-1926) who succeeded his father as chief engineer and commanded engineering and construction until completion.

Emily W. Roebling (1843-1903) who saw that her husband's orders were faithfully executed when he became disabled.

Deutschland Johann August Röbling 1806-1869

145

Above: In 1983, the U.S. Postal Service issued a stamp in honor of Roebling during the celebration of the German-American Tricentennial. [From the collection of Manfred Schnetzer.]

To the right: German stamp issued by the Federal Republic of Germany in 2006 on the occasion of the 200th anniversary of Roebling's birth [Courtesy of Michael Weiss]

pension Bridge was the only crossing point over the Ohio River for its more than eight hundred miles. In 1955, the State of Kentucky replaced the wooden floor with metal grating, which provides a distinctive sound when driving over it, causing it often to be referred to as the "singing" bridge. Automobiles paid a toll until 1963, when the Brent Spence Bridge (Interstate 75) was opened and the toll houses were removed. In 1968, Kentucky's Department of Highways declared the bridge obsolete, and subject to closing, since there were other bridges across the Ohio River. Although it was true that there were other bridges, none of them was like the Roebling Suspension Bridge, and none had the same special significance, and the bridge remained open.

The suspension bridge was placed on the National Register of Historic Places in 1975, was given a fresh coat of blue paint for the celebration of the American Bicentennial in 1976, and was named an engineering landmark in 1983. Although it was originally known as the

The Roebling Suspension Bridge illuminated with lights. [Photograph used with permission from Stevie Publishing Co.]

Covington–Cincinnati Suspension Bridge, it was officially renamed the John A. Roebling Suspension Bridge on 27 June 1983. A new coat of paint prompted one author to write an article entitled, "What Would Roebling Think?" The article concluded, "The graceful old bridge with the new blue color remains as a memorial to a genius engineer with a passion for bridge building. 'Mach's gut,' his friends had told him. John Roebling did."[30]

On 3 September 1984, the Roebling Bridge was illuminated with lights in honor of the memory of Julia Langsam (1905–84), who had served as president of the Covington–Cincinnati Bridge Committee, and who had worked to fly flags and illuminate the bridge for the 1976 Bicentennial. Her son, Walter E. Langsam, stated, "It was very much her project. It all began with patriotism. She wanted flags on the Bridge for the Bicentennial celebration." He also commented that, "architecturally, it's very, very powerful, almost like the great Roman triumphal arches. It's terribly monumental, some parts of it. Still, there are the very light, graceful cables at the same time. It is strong, yet light and delicate."[31]

The original spheres and Greek crosses had been replaced with steel domes in 1897. In March 1992, the golden spheres and crosses on top of the bridge towers were restored—for the first time in ninety-five years. The steel was coated with glue and covered with sheets of 23-karat gold, each two molecules thick. The spheres were lifted into place from the Cincinnati side of the river by helicopter. Other restoration work included masonry work on the towers, metal cleaning, painting, some structural steel repairs, and some repair work on the Cincinnati approach; the total cost of the restoration work cost $10 million.

In late March 2007, the *Cincinnati Post* announced that the Suspension Bridge was ready to reopen after having been closed since November 2006 for a series of structural repairs that included upgrading the steel grid deck and support beams and replacing electrical wiring. Total costs for the refurbishment of the bridge came to $3.1 million, but that was not the only treatment planned for the bridge, which had not been painted since 1980. Plans also called for stage two of the facelift with a new coat of paint.[32]

By the twentieth century, the John A. Roebling Suspension Bridge had become one of the most photographed sites in the region, as well as the veritable symbol of the region, just as the Fountain Square has become a symbol for Cincinnati, and the Glockenspiel has for Covington. It is undoubtedly the most beautiful bridge in the entire Ohio River Valley.

Gerard Roberto, a past president of the Covington–Cincinnati Suspension Bridge Committee, has rightly observed that the bridge serves as a local and national landmark.[33] In addition, the bridge also functions as a "key and symbolic link" to the region's history and heritage and has been called a "national treasure."[34] Hopefully, this work will contribute to an appreciation of the bridge, its creator, and the important role they both play in the material culture and historical identity of the region. It certainly is one of the finest, if not the finest, gem in the crown of the Queen City of the West.

Roebling Suspension Bridge on the Ohio River showing the steel domes that replaced the original spheres and Greek crosses in 1897. In 1992, after nearly a century, the original golden spheres and crosses were restored to the bridge towers. [Photograph courtesy of Little Miami Publishing.]

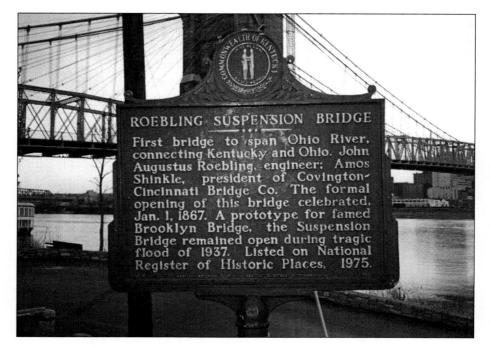

ROEBLING SUSPENSION BRIDGE

First bridge to span Ohio River, connecting Kentucky and Ohio. John Augustus Roebling, engineer; Amos Shinkle, president of Covington-Cincinnati Bridge Co. The formal opening of this bridge celebrated, Jan. 1, 1867. A prototype for famed Brooklyn Bridge, the Suspension Bridge remained open during tragic flood of 1937. Listed on National Register of Historic Places, 1975.

Historical Marker on the Covington side of the Suspension Bridge [from the author's collection].

BRIDGE **STATISTICS**

WEIGHTS AND MEASURES
- Main span from center to center of towers—1,057 feet
- Total length between abutments—1,619 feet
- Length of Cincinnati approach from Front Street to abutment—341 feet
- Length of Covington approach from Second Street to abutment—292 feet
- Total length (including approaches) —2,252 feet
- Greatest weight resting upon the foundation of each tower—32,000 tons
- Area of each foundation—8,250 sq. feet
- Pressure upon each foot—3.88 tons
- Cubic content of masonry of each tower—400,000 feet

RECEIPTS AND EXPENDITURES
- Receipts and expenditures to 28 February 1867—$1,768,821.77

Historical marker located at Sixth and Hornberger avenues in Roebling, New Jersey, which is only a few blocks from the Roebling main Gate Museum. [Photograph courtesy of Jerry McFeeters and Judith Lapp].

ROEBLING'S **SUSPENSION BRIDGES**
1844–1952

John A. Roebling

1844	Allegheny Aqueduct Bridge	Pittsburgh, Pa.	162' spans
1846	Smithfield Bridge	Pittsburgh, Pa.	188' spans
1848	Lackawaxen Aqueduct Bridge	Northeast Pennsylvania	120' spans
1848	Delaware Aqueduct Bridge	Northeast Pennsylvania	134' spans
1850	High Falls Aqueduct Bridge	Southeast New York	145' spans
1850	Neversink Aqueduct Bridge	Southeast New York	
1854	Niagara River Bridge	New York–Canada	821' span
1859	Allegheny Bridge	Pittsburgh, Pa.	344' spans
1867	Cincinnati–Covington Bridge	Ohio–Kentucky	1056' span
1883	Brooklyn Bridge	NYC–Brooklyn, NY	1595' span

John A. Roebling Son's Company

1903	Williamsburg Bridge	NYC–Brooklyn, NY	1600' span
1909	Manhattan Bridge	NYC–Brooklyn, NY	1470' span
1916	Parkersburg Bridge	West Virginia–Ohio	775' span
1922	Roundout Creek Bridge	New York State	705' span
1924	Bear Mountain Bridge	NY State (Hudson River)	1632' span
1931	Maysville Bridge	Kentucky–Ohio	1060' span
1931	Maumee River Bridge	Northwest Ohio	785' span
1931	St. John's Bridge	Portland, Oregon	1207' span
1931	Grand Mere Bridge	Canada	1207' span
1931	Dome Bridge	Arizona	802' span
1932	George Washington Bridge	New York–New Jersey	3500' span
1933	San Rafael Bridge	California	451' span
1937	Golden Gate Bridge	San Francisco, CA	4200' span
1940	Tacoma Narrows Bridge	Tacoma, Washington	2800' span
1943	Peace River Bridge	Alcan Highway Canada	2000' span
1947	Adams Suwanee River Bridge	Northern Florida	
1952	San Marcos Bridge	San Salvador, El Salvador	

THE COVINGTON–CINCINNATI
SUSPENSION BRIDGE COMMITTEE

Continued interest in the suspension bridge is best reflected in the Covington–Cincinnati Suspension Bridge Committee, which was established in 1975 when the bridge was named a National Historic Landmark. According to its Web site, the committee is "a citizens group dedicated to the preservation and enhancement of the John A. Roebling Bridge. The group is responsible for funding the installation and maintenance of the Bridge Beautification Lighting and Flags." Its Web site further notes that "led by Ed Wimmer Sr., the committee incorporated in Kentucky as a non-profit organization dedicated to the general purpose of the flag program. Led by Julia Langsam and Ben Bernstein, the corporate charter was amended in 1984 to include the installation, maintenance and continued operation of the bridge beautification lighting." The committee rightfully notes that "few communities can claim a national historic landmark as their distinguishing symbol. Since 1867, the image of 'Our Suspension Bridge' has been the corporate symbol of local government and business interests, displayed in mass media and rendered in art form adorning the walls of homes and businesses. Buildings and stadiums vie to be its neighbor and city parks and river front developers claim it as their centerpiece icon. Today, more than ever, this bridge speaks loudly in favor of the regional community it unites and the citizens who boast of its possession."

Aside from fulfilling its basic mission, the committee sponsors events, such as the RoeblingFest in June, and takes place in an ideal location on the banks of the Ohio River at the foot of the Suspension Bridge on the Kentucky side of the river just west of Roebling's statue. It features a variety of booths and historical exhibits and artifacts pertaining to the bridge, as well as historical tours of the bridge itself.

After its founder, Ed Wimmer, Sr., the following have served as presidents of the Covington–Cincinnati Suspension Bridge Committee:

Pat Dammert, 1976–77
Pat Coleman, 1977–80
Ed Wimmer, Jr., 1980–82
Julia Langsam, 1982–84
Robert Carter, 1984–85
Robert Wolnitzek, 1985–87
Adrath Abel, 1987–89
Mark Wolnitzek, 1989–91
Bob Armstrong, 1991–93
Ray Noll, 1993–94
Chris Payne, 1994–96
Carl Abel, 1997–99
Gerry Roberto, 1999–2001
John Banner, 2001–03
Ralph Wolff, 2003–05
Anne Huddleston, 2005–07

For further information, as well as a membership form, see the committee's Web site at www.roeblingbridge.com.

The Covington–Cincinnati Suspension Bridge Committee
P.O. Box 17777
Covington, KY 41017-0777

ROEBLING
—H. A. Rattermann

For those who can read German, the following pages are from an article about Roebling, by H. A. Rattermann, and appeared in his *Gesammelte ausgewählte Werke* (Cincinnati: Selbstverlag des Verfassers, 1911).

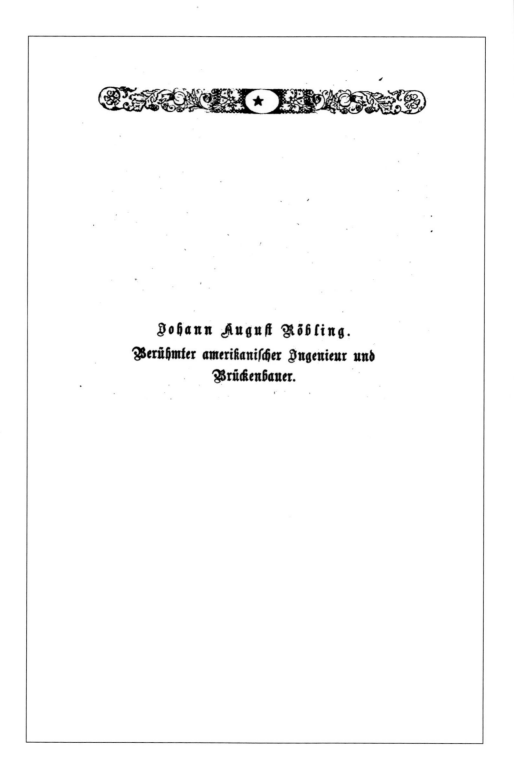

Johann August Röbling.

Berühmter amerikanischer Ingenieur und Brückenbauer.

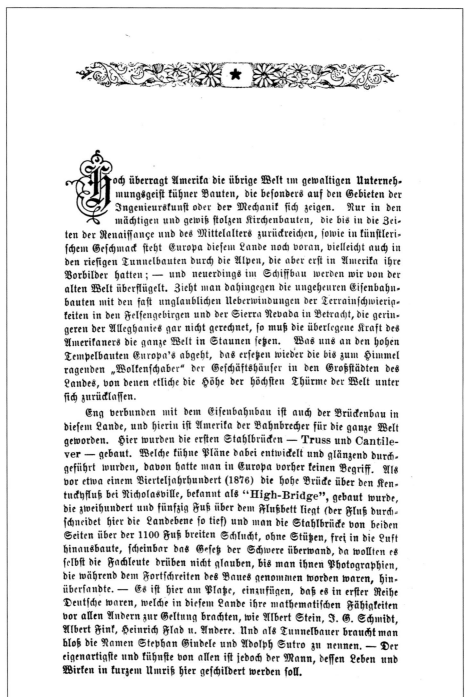

Hoch überragt Amerika die übrige Welt im gewaltigen Unternehmungsgeist kühner Bauten, die besonders auf den Gebieten der Ingenieurskunst oder der Mechanik sich zeigen. Nur in den mächtigen und gewiß stolzen Kirchenbauten, die bis in die Zeiten der Renaissançe und des Mittelalters zurückreichen, sowie in künstlerischem Geschmack steht Europa diesem Lande noch voran, vielleicht auch in den riesigen Tunnelbauten durch die Alpen, die aber erst in Amerika ihre Vorbilder hatten; — und neuerdings im Schiffbau werden wir von der alten Welt überflügelt. Zieht man dahingegen die ungeheuren Eisenbahnbauten mit den fast unglaublichen Ueberwindungen der Terrainschwierigkeiten in den Felsengebirgen und der Sierra Nevada in Betracht, die geringeren der Alleghanies gar nicht gerechnet, so muß die überlegene Kraft des Amerikaners die ganze Welt in Staunen setzen. Was uns an den hohen Tempelbauten Europa's abgeht, das ersetzen wieder die bis zum Himmel ragenden „Wolkenschaber" der Geschäftshäuser in den Großstädten des Landes, von denen etliche die Höhe der höchsten Thürme der Welt unter sich zurücklassen.

Eng verbunden mit dem Eisenbahnbau ist auch der Brückenbau in diesem Lande, und hierin ist Amerika der Bahnbrecher für die ganze Welt geworden. Hier wurden die ersten Stahlbrücken — Truss und Cantilever — gebaut. Welche kühne Pläne dabei entwickelt und glänzend durchgeführt wurden, davon hatte man in Europa vorher keinen Begriff. Als vor etwa einem Vierteljahrhundert (1876) die hohe Brücke über den Kentuckyfluß bei Nicholasville, bekannt als "High-Bridge", gebaut wurde, die zweihundert und fünfzig Fuß über dem Flußbett liegt (der Fluß durchschneidet hier die Landebene so tief) und man die Stahlbrücke von beiden Seiten über der 1100 Fuß breiten Schlucht, ohne Stützen, frei in die Luft hinausbaute, scheinbar das Gesetz der Schwere überwand, da wollten es selbst die Fachleute drüben nicht glauben, bis man ihnen Photographien, die während dem Fortschreiten des Baues genommen worden waren, hinübersandte. — Es ist hier am Platze, einzufügen, daß es in erster Reihe Deutsche waren, welche in diesem Lande ihre mathematischen Fähigkeiten vor allen Andern zur Geltung brachten, wie Albert Stein, J. G. Schmidt, Albert Fink, Heinrich Flad u. Andere. Und als Tunnelbauer braucht man bloß die Namen Stephan Gindele und Adolph Sutro zu nennen. — Der eigenartigste und kühnste von allen ist jedoch der Mann, dessen Leben und Wirken in kurzem Umriß hier geschildert werden soll.

422

Johann August Röbling, einer der erſten Ingenieure der Neuzeit, der als Brückenbauer Robert Stephenſon und die Ingenieure Brunel (Vater und Sohn) weit überflügelt hat, wurde am 12. Juni 1806 zu Mühlhauſen in Thüringen geboren, beſuchte das dortige Gymnaſium und bildete ſich auf der Realſchule in Erfurt und dann am Politechnikum zu Berlin zum Ingenieur aus. In Berlin beſuchte er zu gleicher Zeit die Vorleſungen Hegel's, da ihn, außer dem Fachſtudium, auch die Philoſophie anzog. So wiſſenſchaftlich und techniſch ausgebildet, erhielt er eine An= ſtellung von der preußiſchen Regierung beim Wegebau in Weſtfalen (1827), welche Stellung er jedoch nach zwei Jahren wieder aufgab, um eine Buch= druckerei und Buchhandlung in Eſchwege zu übernehmen. *) Hier wurde Röbling mit einem proteſtantiſchen Prediger, J. A. Etzler, bekannt, der ſich viel mit phyſikaliſchen Problemen abgab und ſpäter in Amerika Vorleſun= gen über die Nutzbarmachung der Naturkräfte hielt, dadurch das unterneh= mungsluſtige Volk und ſelbſt den Kongreß eine zeitlang in athemloſer Spannung haltend.

Um dieſe Zeit (1831) war die Kolonifationsbewegung nach Amerika in Deutſchland lebendig im Gange, und auch in Thüringen wurde eine Kolonifationsgeſellſchaft in's Leben gerufen, die ihren Hauptſitz in Mühl= hauſen hatte. Der Anreger und Leiter war Friedrich Chriſtoph von Dach= röden, und an dieſen ſchloſſen ſich als Führer die Herren Johann A. Etzler, Heinrich Harſeim (Kaufmann in Eiſenach) und unſer J. A. Röbling, viel= leicht auch der ſpätere St. Louiſer Kaufmann Emil Angelrodt an. Sie veröffentlichten im Röbling'ſchen Verlag eine Druckſchrift: „Allgemeine Anſicht der Vereinigten Staaten von Nordamerika", Eſchwege, 1831, in welcher ſie den Plan einer Anſiedlung im Süden des Landes (Florida, Louiſiana oder Arkanſas) auseinanderſetzten. Röbling und Etzler wurden als Kundſchafter vorausgeſchickt (1831) und ſie wählten Land in Beaver County, Pennſylvanien, in der Nähe von Rapp's „Economy" aus. Im darauffolgenden Jahr ſchiffte ſich, wie Brauns (S. 295) berichtet, ihre Ge= ſellſchaft zu Bremen ein; ob ſich aber alle in Pennſylvanien niederließen, wird nicht mitgetheilt.

In dieſer deutſchen Kolonie, „Neu Sachſen", ließen ſich Röbling und Etzler als Landwirthe nieder, allein Etzler erlahmte bald und zog nach Pittsburg, wo er mit J. G. Backofen in Theilhaberſchaft eine deutſche Zei= tung in's Leben rief, den „Pittsburg Beobachter" (1834), der ein Jahr ſpäter an Schmidt und Backofen überging und in „Adler des Weſtens" umgetauft wurde, worauf Etzler nach Cincinnati ging und ſeine Vorträge über die Nutzbarmachung der Naturkräfte begann. Ein Büchlein Etzler's,

---

 *) Siehe Dr. E. L. Brauns: „Amerika und die moderne Völkerwanderung" — Potsdam 1833, Seite 294.

im Verlag von Schmidt und Storch, Cincinnati, 1835, gibt Auffchluß über seine Theorien : Auffangen der Sonnenstrahlen durch mächtige Brennspiegel, und Auffspeicherung der Wärme für den Winter. Anwendung des Windes zum Pflügen mittelst riefiger Drachen. Künstliche Infeln welche im Ozean verankert und zu Sommervillas benutzt werden sollten 2c. Auch die Elektrizität wollte er benutzen zur Transmittirung von Botschaften frei durch die Luft. Dadurch ist Etzler der Vorläufer Marconi's gewesen, allein seine Theorien ließen sich nicht praktisch verwerthen.

Auch Röbling wurde des Farmlebens nach einigen Jahren überdrüffig und sehnte sich wieder zu seinem Beruf eines praktischen Ingenieurs. Im Herbst 1837 war er jedoch noch auf seiner Farm und betheiligte sich als Abgeordneter von Beaver County, Pa., und ebenfalls an der zweiten deutschen Konvention (1838), aber nicht mehr an der dritten vom Jahre 1839.

Um diese Zeit herrschte eine große Thätigkeit im Bau von Kanälen in den Vereinigten Staaten. Der Staat New York hatte seinen großen Erie-Kanal von Troy nach Buffalo im Jahre 1832 vollendet ; Ohio den Kanalbau von Cleveland nach Portsmouth am Ohiofluffe, den „Ohio Kanal", im Jahre 1833 kompletirt und der „Miami Kanal," von Toledo nach Cincinnati, ging 1835 seiner Vollendung entgegen. Dieses lenkte den Verkehr vom Often nach dem Westen, von Philadelphia und Baltimore ab, nach New York hin. Der Bau von Eisenbahnen war damals noch in den Erstlingsstadien und so versuchten denn die Kapitalisten von Philadelphia und Baltimore es ebenfalls mit dem Kanalbau. Diesem setzten indeffen die Alleghanygebirge eine unüberfteigliche Barriere in den Weg. Man zerbrach sich die Köpfe darüber, wie man dieses Hinderniß bewältigen könne, aber kein Ingenieur fand den Ausweg.

Da war es Röbling, der das Mittel fand, um diese Schwierigkeit zu überwinden. Er trat mit der Kanal-Gesellschaft in Verbindung und erbot sich, mittelst Konstruktion einer Anzahl von Schiefebene Bahnen die Kanalboote, welche zu dem Zweck in vier auseinandernehmbaren Theilen gebaut werden müßten, über die Berge zu heben und jenseits derselben wieder ebenso herunterzulaffen. Das war eine ganz neue Idee und leitete Röbling zu dem Unternehmen der Fabrikation von Drahtfeilen, die später das Fundament zu seinen Draht-Hängebrücken wurden. Der Kanal wurde darauf bis Holydaysburg gebaut, dann durch ein Syftem von acht oder zehn Schiefebene Bahnen die Boote über die Berge geführt und jenseits derselben bei Johnstown wieder in den von Pittsburg dahingebauten Kanal gebracht. Dieses Werk, mittelst Drahtfeilen den Weg über die Pennfylvania Portage zu vermitteln, wurde im Jahre 1842 vollendet. Der Ruf Röbling's war damit gesichert.

Zunächst galt es, noch eine Schwierigkeit zu überwinden, nämlich den Bau einer Brücke über den Alleghanyfluß nach Pittsburg hinein. Die

424

Brückenbauer erklärten das Problem für unlösbar. Da erbot sich Röbling im Jahre 1844 den Kanal-Aquedukt über den Fluß an Drahtseile zu hängen, ein Werk, welches zur Zeit das größte Aufsehen erregte und seinen Ruf als Ingenieur fest begründete. Zunächst erbaute Röbling dann die schöne Monongahela Hängebrücke bei Pittsburg, fünfzehnhundert Fuß lang mit acht Spannungen.

Er war nun in das rechte Fahrwasser gekommen und kräftig segelte er mit dem Strom weiter. Dabei wuchsen seine Pläne. Schon 1846 kam er nach Cincinnati und machte Herrn Robert Bowler, dem Haupteigenthümer der Kentucky Central Eisenbahn, den Vorschlag, eine Hängebrücke über den Kentuckyfluß bei Nicholasville zu bauen, die auch in Angriff genommen wurde. Die Thürme an den beiden Seiten des Flusses sind vollendet, allein da machte die Kompagnie Bankerott und der Ausbau unterblieb. Zur selben Zeit wurde auch von Röbling den Cincinnatier Kapitalisten der Vorschlag gemacht, die Hängebrücke über den Ohiofluß zu bauen, allein die langwierigen Verhandlungen in den Gesetzgebungen von Ohio und Kentucky, wo allerhand Einwendungen erhoben und die Gesellschaft, welche von Röbling in's Leben gerufen worden war, so lange hinausgehalten ward, bis der Bau der Brücke in Angriff genommen werden konnte. Die Hauptgegner waren die Dampfbootleute, welche glaubten, daß die Schifffahrt auf dem Flusse dadurch gehemmt werden würde. In der Ohioer Legislatur machte sich die Ansicht geltend, daß eine so große Spannung unmöglich sei, und in Kentucky gab man vor, daß die Brücke für eine Flucht der Neger aus der Sklaverei gefahrdrohend werden könnte. Beide Einwände wurden später gehoben.

Inzwischen trat Röbling mit der Pennsylvania Eisenbahn Gesellschaft in Unterhandlung behufs Bau einer Brücke über den Delawarefluß bei Trenton, N. J., und errichtete im Jahre 1848 eine Drahtseil Fabrik in Trenton, die nach und nach in großartiger Weise erweitert wurde. Damals erhielt er den Kontrakt für den Bau der berühmten Hängebrücke über den Niagarafluß etwa eine Meile unterhalb den Fällen. Es ist ein wundervoller Bau von 800 Fuß Spannung über das in 200 Fuß tiefer Schlucht fort tosende Wasser. Die Brücke ist ein sog. Zweidecker. Ueber den Weg für Wagen und Fußgänger befindet sich ein zweites Stockwerk für die Eisenbahn. Das Werk ist von höchster Eleganz und hat sich seine Festigkeit schon über ein halbes Jahrhundert bewährt. Es ist ein majestätischer Anblick, wenn man auf der Eisenbahn hoch über die brausenden Rapids dahinfährt und dicht daneben den donnernden Sturz des gewaltigen Katarakts sieht. Man fühlt unwillkürlich die Größe und Macht des Menschengeistes, welcher dieses Hinderniß überwinden konnte. Die Brücke wurde im Jahre 1854 vollendet.

Die Sicherheit, mit welcher die Schnellzüge der Eisenbahn in größter Geschwindigkeit und die schwersten Frachtzüge über die Brücke fuhren, löste

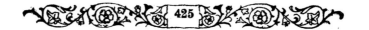

auch den letzten Zweifel über die mathematisch genauen Berechnungen Röbling's und sein Ruf als der erste Ingenieur der Zeit erscholl nicht nur über das ganze Land sondern auch nach Europa hin. Im Jahre 1855 kam er dann wieder nach Cincinnati zurück, wo er auf's Neue den Plan zum Bau der Hängebrücke über den Ohiofluß in Angriff nahm. Der Kongreß sowohl wie die Legislatur von Kentucky hatten jetzt auch die Einwendungen gegen den Bau der Brücke gehoben und der von Röbling gegründeten Brückengesellschaft wurde in Kentucky der nöthige Freibrief gewährt. Im Jahre 1856 begann man mit den Ausgrabungen und den Bau der Fundamente auf beiden Seiten des Flusses. Dieser kolossale Unterbau verschlang weit mehr als die Hälfte der Baukosten der ganzen Brücke. Er beschäftigte den Meister fünf Jahre lang, während welcher Zeit er, mit Ausnahme einiger Wintermonate, sich in Cincinnati aufhielt und sich gern am Stammtisch in dem Gloßner'schen Lokal betheiligte.

Hier lernte auch der Verfasser dieses ihn als einen geistreichen Gesellschafter kennen; und selbstverständlich kamen der Brückenbau und die technischen Schwierigkeiten desselben häufig zur Sprache. Natürlich wurden mancherlei Fragen dabei aufgeworfen, besonders in Bezug auf die Tragkraft der Brücke. Die Tragkraft sei es nicht, meinte Röbling, welche zu überwinden sei, sondern die Expansion des Metalls im Sommer und die Kontraktion desselben im Winter. Daraus ergebe sich, daß ohne Vorbeugung der Druck auf die Thürme nicht immer senkrecht fallen dürfte, allein sein Plan überwinde auch diese Schwierigkeit, indem die Kabel über mächtige Rollen liefen, die er oben auf den Thürmen anbringen werde, wodurch diese Wechsel ausgeglichen würden; zunächst betraf es aber auch die sichere Verankerung der Kabel an ihren Enden. Die dritte Einwendung galt einer Vorkehrung gegen die mächtigen Windstürme, welche der ungeheuren Spannung der Brücke drohen möchten. So bedeutend diese Gefahr auch sein dürfte, entgegnete Röbling, ich begegne sie damit, daß die beiden Kabel in der Mitte der Brücke näher zusammenlaufen, wodurch jedes Schwanken des Baues verhindert wird. Durch diese Erklärungen Röblings wurden wir von der Gründlichkeit seiner Pläne überzeugt, und die Deutschen von Cincinnati zählten bald zu den Hauptbefürwortern des Unternehmens und Vertheidigern von Röbling's Ideen, wenn auch untergeordnet nur als geldliefernde Aktionäre.

Im Frühjahr 1861 waren die Thürme ausgebaut, und auch die Drahtseile waren fertig und nach Cincianati gebracht worden, um über den Fluß gesponnen zu werden. Da brach der leidige Bürgerkrieg aus und brachte das Werk auf kurze Zeit in's Stocken. Im Frühjahr 1864 wurde jedoch die Vollendung wieder in Angriff genommen und im Herbst desselben Jahres waren die Drähte gezogen, die mächtigen Kabel umsponnen und der Boden der Brücke gelegt. Am Neujahrstag 1865 wurden etwa

426

tausend Personen eingeladen zum ersten Mal die Brücke zu überschreiten, bei welcher Gelegenheit Röbling auf der Brücke eine Ansprache an die Versammlung hielt, in welcher er die Dauerhaftigkeit des gewaltigen Baues auseinandersetzte und zugleich mittheilte, daß er noch einen weit größeren Brückenbau im Plan vollendet und bereits in Angriff genommen habe, „das Ideal meines Lebens", wie er sich ausdrückte, den Bau der Brücke über den Eastriver, die Städte New York und Brooklyn verbindend. — Die Dimensionen dieser ebenso schönen als eleganten Brücke, welche Cincinnati mit dem gegenüberliegenden Covington verbindet, sind: die Hauptspannung zwischen den beiden Thürmen 1128 Fuß; die ganze Länge derselben, nach dem Ausbau des Aufgangs in Cincinnati bis zur Zweiten Straße zirka 2800 Fuß; die Höhe der Brücke in der Mitte des Flusses über den normalen Wasserspiegel des Ohio 110 Fuß, so daß außer den seltenen Hochfluthen die größten Dampfer unter der Brücke hinwegfahren können; die Höhe der beiden Pfeiler über den Fundamenten ist 178 Fuß und mit den 72 Fuß tiefen Grundmauern 250 Fuß.

Sein letztes großes Werk war, wie gesagt, der Entwurf der kolossalen Brücke über den Eastriver, New York mit Brooklyn verbindend. Röbling hatte sich über ein Jahrzent mit diesem Plan getragen, und nachdem Alles durchdacht, Alles berechnet und im Plan vollendet war, nachdem die zahllosen Schwierigkeiten und die starke Opposition nur durch den eisernen Willen eines einzigen Mannes besiegt waren, da mußte in dem ersten Moment des praktischen Beginnens der große Baumeister hinweggerafft werden. Ein Balken quetschte ihm den linken Fuß. Es mußten ihm vier Zehen abgenommen werden. Die Wunde fing an zu heilen und alle Gefahr schien beseitigt, als sich die Mundsperre einstellte und den kräftigen Mann nach langem Leiden hinwegraffte. Er starb mit philosophischer Ruhe am 20. Juni 1869. Das große Werk, dessen Bau bereits im Fortschreiten begriffen war, wurde von seinen Söhnen, die schon lange mit dem Vater gedacht und gearbeitet hatten, unter Leitung des ältesten derselben, Washington A. Röbling, weitergeführt und zur Vollendung gebracht. Zwar wollte man im Herbst 1882 Washington Röbling, der bis dahin der Ober-Ingenieur des Baues war, sich aber bei den Fundamentbauten ein Leiden zugezogen hatte, aus seiner Stellung verdrängen, allein dazu ist es infolge zahlreicher Proteste, die von allen Seiten erhoben wurden, nicht gekommen, und er konnte noch vier Jahre später die Vollendung des gewaltigen Baues von seinem Krankenzimmer in Brooklyn sehen.

Diese Brücke mißt ihrer ganzen Länge nach über sechstausend Fuß, mit einer Hauptspannung zwischen den Pfeilern von 1600 Fuß. Ihre Höhe ist daraus zu ermessen, daß die größten Schiffe mit den höchsten Masten zu allen Zeiten unter derselben hinwegsegeln können. Die Brücke wird von vier mächtigen Kabeln gehoben, die je 18 Zoll im Durchesser haben.

427

(Auch der Cincinnatier Brücke sind seitdem noch zwei weitere Kabel hinzugefügt worden, die wie die alten je 13 Zoll Durchmesser haben.) Gegenwärtig (1903) ist noch der Bau einer zweiten Brücke über den Eastriver nach dem Röbling'schen System im Fortschritt begriffen und geht der Vollendung entgegen. Dieselbe wird eine Spannung zwischen den Thürmen von 1750 Fuß erhalten. — Was sind alle Weltwunder der alten Zeit gegen diese Wunderwerke unseres deutschen Landsmannes? — — —

Aber nicht nur als einer der hervorragendsten seines Faches, sondern auch als Privatmann war er einer der edelsten und besten. Der Geistliche, der unter dem Andrang von mehr als fünftausend Personen zu Trenton das Andenken des Verstorbenen bei dessen Bestattung feierte, sagte unter Anderem: „In ihm hat Trenton einen seiner besten Bürger, die Armen haben einen ihrer größten Wohlthäter und die Welt einen erleuchteten Geist verloren." Er war der alleinige Erhalter des Waisenhauses und der liberale Unterstützer anderer Wohlthätigkeits Anstalten. Er hinterließ ein sehr großes Vermögen, und in seinem Testament setzte er bedeutende Summen für wohlthätige Institute aus, gleichfalls für die Gründung einer polytechnischen Schule in Trenton bestimmte er ein namhaftes Vermächtniß.

Röbling war ein außerordentlicher Mann. Eine mächtige, hohe Stirn überwölbte seine mit energischen Brauen bedeckten, etwas tiefliegenden Augen. Man hielt ihn beim ersten Begegnen für einen finsteren, unzugänglichen Karakter, aber wer in seine Gesellschaft öfters kam, wurde leicht vom Gegentheil überzeugt. Er war dann nicht gerade redselig, aber ließ doch das gesellige Temperament gern durchblicken. Dr. Schmöle in seiner Karakterisirung der Theilnehmer an der ersten deutschen Konvention zu Pittsburg (1837), nennt Röbling den „philosophisch Strebenden". Philosophie war in Gesprächen seine liebste Unterhaltung. Verfasser dieses erinnert sich von einem öfteren Zusammentreffen Röbling's mit dem ebenfalls Hegelianer, Peter Kauffmann, aus Canton, Ohio, in der Weinstube von Nikolaus Schmitt im Jahre 1858. Kauffmann hatte damals in Cincinnati sein Buch im Druck: „Tempel der Wahrheit", eine ziemlich phantastische Darstellung der Natur- und Weltbegriffe, ganz im Hegel'schen Geiste gedacht, nur noch höher auf die Spitze getrieben als Hegel. Dabei glaubte er, seine dunkelen Begriffe für den allgemeinen Volksgebrauch erläutert zu haben. Ich besitze das Buch, habe es aber nie mit Verständniß durchlesen können. Diese beiden Männer kamen damals häufig zusammen, wobei dann Philosophie das Gespräch bildete. Dann leuchteten auch Röbling's Augen auf, aus denen der Genius blitzte. In seinen Gesprächen war er doch bedachtsamer, als sein philosophischer Kollege, und überstieg sich nicht, wie dieser, oft in's Absurde. Röbling war, obschon er die hegel'schen Vorlesungen gehört hatte, mehr Kantianer geblieben. Auch in seinen Un-

terhaltungen war er stets bedächtig und überhastete sich nie, wenn auch sein Gegner oft die bizarrsten Argumente vom Stapel ließ. — Selbstverständlich sprach Röbling auch gern von seinen Ideen, allein dann waren wir Andern meist stille Zuhörer der gewaltigen von ihm wohlüberdachten Pläne. Die Erklärung seiner himmelstürmenden Projekte schien uns häufig als unausführbar, allein die Folge hat gelehrt, daß Röbling Recht hatte.

Er war von imponirender Gestalt, groß und breitschultrig gebaut. Mund und Nase zeugten von Entschlossenheit, Kraft und Kühnheit. Beim ersten Anblick vermuthete man nicht seine Güte des Herzens, die große Bescheidenheit, die ihn zierte, sowie seine Leutseligkeit im Umgang, aber wenn man öfters mit ihm verkehrte, traten auch diese Eigenschaften lebendig zu Tage. In einem Nachruf, den ihm die Aktionäre der Covington und Cincinnatier Brückenkompagnie widmeten, heißt es unter Anderem: „In dem Zweige der Kunst, welcher er den größten Theil seines Lebens gewidmet hat, war Niemand seines Gleichen. Während er alle Autoritäten seines Faches achtete, ließ er sich dennoch durch dieselben nicht binden. Ihm gebührt der Ruhm, beide Ströme (Niagara und Ohio) mit Sicherheit überbrückt und die Schifffahrts-Kanäle frei und ungesperrt gelassen zu haben. Sein letztes vollendetes großes Werk in unserer Stadt wird für Jahrhunderte stehen als ein öffentlicher Segen und als ein Denkmal seines umfassenden Genius."

*   *   *

August Röbling.

Herrlicher Geist! du bezwangest den Himmel, das Meer und die Erde!
     Fest in der Felsen Schooß grubst du den eisernen Zahn,
Spanntest dein fliegend Gewebe dann über den Strom durch die Luft hin,
     Brausenden Stürmen zum Trotz hoch in den Aether gebannt.
Und Millionen nun stürmen beflügelten Fußes hinüber,
     Und ein jeglicher Schritt tönt dir unsterblichen Ruhm!

BRIDGE OVER THE OHIO
—B. Drake and E. D. Mansfield

incinnati in 1826 (Cincinnati: Printed by Morgan, Lodge, and Fisher, 1827), by B. Drake and E. D. Mansfield, reminds us that the idea of building a bridge across the Ohio River had long been discussed. The authors write the following:

This subject has been one of much speculation for several years past. Its importance is perhaps not less apparent, than the practicability of its execution. The scarcity of capital among our citizens may delay it for a few more years, but the period is manifestly not remote, when its construction will be undertaken.

The feasibility of throwing a permanent bridge over the Ohio at this place, at an expense which would secure a handsome interest upon the sum required for its accomplishment, is generally admitted, by those practical, calculating men, who have had the subject under consideration, and who have possessed the existing data, from which to draw their conclusions.

The water of the Ohio passes over a bed of limestone rock, which will not only supply the stone, necessary in the construction of the piers and abutments, but also, an admirable foundation for them to rest upon. The distance from the top of the bank at the foot of Broadway, to the top of the bank in Newport or Covington, is 1,630 feet, or about 543 yards. What is termed the channel of the river lies near the north shore; its south edge is 435 feet distant from the wall at the foot of Broadway. There is in this channel a gradual descent from the north to the south edge; the distance from one to the other being about 225 feet. Should this space be thought too great to exist with safety between the piers, an intermediate one may readily be constructed in the channel, the greatest depth at low water not exceeding 12 feet. The whole distance across the river would require 8 or 9 piers, besides the abutments on either bank. From the foot of Broadway, a bridge would strike the Kentucky shore, oppo-site the mouth of Licking. A line drawn from the bank on the Newport shore, until, at a distance of 200 feet from the place of beginning, it should intersect a similar line, from the Covington shore, would indicate the proper point for a pier, on which the main bridge should terminate. From this, branches should be carried to Newport and Covington, thus uniting those two villages with each other, and both with Cincinnati.

Between the shore and the northern edge of the channel, there is, during the high water, an eddy, formed by the steam mill above, over which, the draw may properly be made to admit the passage of steam boats at that stage of the river: at a medium stage, the elevation of the bridge over the main channel of the stream, would be such as to permit the passage of the largest class of boats. Various estimates of the cost of this work have been made, varying in amount from one to two hundred thousand dollars. An architect who has superin-tended the construction of several bridges in the Miami Country, and whose practical skill entitles his opinions to confidence, has recently given this subject some consideration. His estimates of the cost of a bridge, of the length above mentioned, supported by nine stone piers, including breakers above each, to protect them from the ice and drift wood; branching so as to connect New-port and Covington, and secured from the weather by a neat and substantial cover, is $150,000.

How nearly this may approximate to the truth, remains to be determined by more accurate surveys. Should it even cost $200,000, still it is believed, that the tolls would, from the time of its completion, yield a handsome inter-est upon its cost, with a certain prospect of an increase, corresponding to the rapid advances of the city and surrounding country. It is hoped that our public *spirited citizens, will not lose sight of an object so deeply connected with the convenience and ornament of the city. If our own resources at the present moment, are not adequate to the magnitude of the work, it would perhaps, be no difficult matter to put in requisition some of the surplus capital of our eastern brethren, to aid in its early accomplishment.*

NOTES

Chapter One

1. See Georg Wilhelm Friedrich Hegel, *The Philosophy of History* (New York: Wiley, 1900), 86. For an introductory selection of Hegel's works, see Kuno Francke, ed., *The German Classics of the Nineteenth and Twentieth Centuries* vol. 7 (New York: German Publication Society, 1913–1914).
2. For an English-language translation of Duden's influential book, see Gottfried Duden, *Report on a Journey to the Western States of North America: And a Stay of Several Years Along the Missouri (during the Years 1824, '25, '26, and 1827)* ed. James W. Goodrich et al. (Columbia, Missouri: State Historical Society of Missouri, 1980).
3. The quotes here and in the following paragraphs by Roebling regarding his book are drawn from John A. Roebling, "Opportunities for Immigrants in Western Pennsylvania in 1831," *Western Pennsylvania Historical Magazine* 18(1935): 73–108
4. For further information on Economy, Pennsylvania, see Karl J. R. Arndt, *George Rapp's Harmony Society*, 1785–1847 (Philadelphia: University of Pennsylvania Press, 1965). See also, Karl J. R. Arndt and Patrick R. Brostowin, "Pragmatism and Prophets: George Rapp and J. A. Roebling versus

J. A. Etzler and Count Leon," *Western Pennsylvania Historical Magazine* 5(1969): 171–98.

5. Ibid. For further information on Saxonburg, Pennsylvania, see Ralph Goldinger, *Historic Saxonburg and its Neighbors* (Saxonburg, Pa.: Saxonburg Historical and Restoration Commission, 1990).

6. See John A. Roebling, *Diary of My Journey from Muehlhausen in Thuringia via Bremen to the United States of North America in the Year 1831, Written for my Friends by Johann Augustus Roebling* trans. with Occasional Notes, from the original by Hamilton Schuyler (Trenton, N.J.: The Roebling Press, 1931).

7. Regarding Reemelin, see Charles Reemelin, *Life of Charles Reemelin in German, Carl Gustav Ruemelin, from 1814–1892* (Cincinnati: Weier & Saiker, 1892).

Chapter Two

8. For information on the Niagara Bridge, see Rudolf Cronau, *German Achievements in America: Rudolf Cronau's Survey History* ed. Don Heinrich Tolzmann (Bowie, Md.: Heritage Books, Inc., 1995), 137–43. Cronau comments on the significance of Roebling as follows: "A complete revolution in bridge-building was brought about during the midst of the 19th century by Johann August Roebling. . . . Soon after his graduation from the Royal Polytechnicum at Berlin he emigrated [*sic*] to the United States and established himself at Saxonburg, Pa. There he developed the manufacture of wire cable for use in bridge construction to a degree unknown before. Bridge building then was, in comparison to its present perfection, in the first stages of development. Suspension bridges were known, but the platforms were hung on heavy chains, the links of which possessed notwithstanding their weight no great holding capacity. Besides, for spans of more than 180 feet they were impracticable. It remained for Roebling to substitute a system of wire-cables, the enormous carrying capacity of which he demonstrated in 1845 in a suspended aqueduct of the Pennsylvania Canal carried across the Monongahela River. This was soon followed by the Monongahela suspension bridge at Pittsburgh and the suspension railway bridge across the Niagara River." Cronau, 137.

See also John A. Roebling, *Report of John A. Roebling to the Presidents and Directors of the Niagara Falls Suspension and Niagara Falls International*

Bridge Companies on the Condition of the Niagara Railway Suspension Bridge, August 1, 1860 (Trenton, N.J.: Murphy & Bechtel, Printers, 1860).

9. See H. A. Rattermann, *Gesammelte ausgewählte Werke* (Cincinnati: Im Selbstverlage, 1911), 12:419–28. For further information on Rattermann, see Mary Edmund Spanheimer, *The German Pioneer Legacy: The Life and Work of Heinrich A. Rattermann* ed. Don Heinrich Tolzmann, New German-American Studies, vol. 26. (Oxford: Peter Lang Pub. Co., 2004).

10. For information on Peter Kaufmann, see Loyd E. Easton, *Hegel's First American Followers* (Athens, Ohio: University Press, 1966); Karl J. R. Arndt, ed., Teutonic *Visions of Social Perfection for Emerson: Verheissung und Erfllung: A Documentary History of Peter Kaufmann's Quest for social Perfection from George Rapp to Ralph Waldo Emerson* (Worcester, Mass.: Harmony Society Press, 1988).

11. See David McCullough, *The Great Bridge* (New York: Simon & Schuster, 1972), 70. McCullough also notes that "Talking in retrospect, Amos Shinkle had nothing but praise for the manner in which it had been built. From an engineering standpoint everything had gone smoothly. Only two lives had been lost during the entire time of its construction, a remarkable safety record. . . . For the Roeblings, he had only the highest admiration, and especially for the redoubtable father . . ." Shinkle also advised those who wanted Roebling to build the Brooklyn Bridge: "He is an extraordinary man and if you people in Brooklyn are wise you will interfere with his views just as little as possible. Give the old man his way and trust him," 70.

On his death in 1892, the Bridge Company issued the following tribute to Shinkle: "He was the unfaltering friend of the Suspension Bridge from its inception to its completion. In the dark days of its construction when friends failed, disappointments multiplied, and among its projectors many wavered and most of them despaired of success—then it was that the strong will of Mr. Shinkle rose to the occasion and with words of fixed determination he declared that the bridge should be built." See Harry S. Stevens, *The Ohio Bridge* (Cincinnati: Ruter Press, 1939), 176.

12. For information about Washington Roebling, see the biographical article in the *Dictionary of American Biography* (New York: Scribner, 1935), 16:89–90.

13. See "John A. Roebling," *Der Deutsche Pionier* 1(1869): 194–201. For a

historical survey of the area's German heritage, see Don Heinrich Tol-zmann, *German Heritage Guide to the Greater Cincinnati Area* (Milford, Ohio: Little Miami Publishing Co., 2003). For information on a Roebling letter in the German-Americana Collection at the University of Cincinnati, see "John Roebling in Stone, Steel, and Ink," *Gathering: A Publication of the Friends of the University of Cincinnati Libraries* 19(2001).

Chapter Three

14. H.J. Ruetenik, *Berühmte deutsche Vorkämpfer für Fortschritt, Freiheit und Friede in Nord-Amerika* (Cleveland, Ohio: Forest City Bookbinding Co., 1899) 251.
15. See *Der Deutsche Pionier*, 1(1869): 195.
16. David B. Steinman, *The Builders of the Bridge: The Story of John Roebling and His Son* (New York: Harcourt, Brace and Company, 1945), 129.
17. Ibid, 21.
18. Ibid, 131.
19. Ibid.
20. Robert M. Vogel as cited in Elizabeth C. Stewart, ed., *Guide to the Roebling Collections at Rensselaer Polytechnic Institute and Rutgers University, With an Introduction by Robert M. Vogel* (Troy, N. Y.: Friends of the Folsom Library, Rensselaer Polytechnic Institute, 1983), ix.
21. Hamilton Schuyler, *The Roeblings: A Century of Engineers, Bridge-Builders and Industrialists: The Story of Three Generations of an Illustrious Family, 1831–1931* (Princeton: Princeton University Press, 1931), 79.
22. Ibid.

Chapter Four

23. See Judith St. George, *The Brooklyn Bridge: They Said It Couldn't Be Done* (New York: Putnam, 1982), 104–05. See also, Samuel W. Green, A *Complete History of the New York and Brooklyn Bridge, from its Conception in 1866 to its Completion in 1883* (New York: S. W. Green's Son, 1883), and also the PBS program *Brooklyn Bridge: A Film by Ken Burns* (Los Angeles, Calif.: Direct Cinema, 1990).
24. See Vivian H. Pemberton, "The Roebling Reponse to Hart Crane's *The Bridge*: A New Letter," *Ohioana Quarterly* 20:4(1977): 155–157.

25. Cronau, *German Achievements*, 134.

26. Dixie M. Golden, ed., *A Golden Gate Bridge Jubilee, 1937–1987, University of Cincinnati, College of Engineering* (Cincinnati: University Publications for the College of Engineering, 1987), n.p.

27. Reginald C. McGrane, *The University of Cincinnati: A Success Story in Urban Higher Education* (New York: Harper & Row, 1963), 287.

28. Steger as cited in Golden, *A Golden Gate Bridge Jubilee*, n.p. For further information on Strauss, see "Joseph Baermann Strauss," *Dictionary of American Biography* suppls. 1–2 (1940). See also, A. William Finke, "Joseph B. Strauss, His Life and Achievements," bachelor's thesis, (University of Cincinnati, 1960). For a more recent history of the University with references to Strauss, see Kevin Grace and Greg Hand, *The University of Cincinnati* (Montgomery, Ala.: Community Communications, 1995), 89, 106–07, 198.

29. See Joseph F. Gastright, "Wilhelm Hildebrand and the 1895 Reconstruction of the Roebling Suspension Bridge," *Northern Kentucky Heritage* 8:1(2000): 1–14. With regard to the reconstruction work Hildebrand stated "It was my belief, based on a thorough examination of all parts of the old bridge, that the structure was in a state of excellent preservation and just as serviceable as when new, and, if the traffic conditions had not been changed, that no repairs, or only immaterial ones, would have been necessary," 5. The other key figure in the reconstruction work was Bradford Shinkle, the son of Amos Shinkle, who served as president of the Covington and Cincinnati Bridge Company, 1892–1909.

30. See Ruth Engelken, "What Would Roebling Think?" *Cincinnati Enquirer Magazine*, March 7, 1976.

31. See Jim Calhoun, "Suspended Delights: John Roebling's Bridge Will Feature A Shining First Once Again," *Cincinnati Enquirer*, August 19, 1984. For information on the restoration work on the bridge, see Lew Moore's, "Suspension Bridge A Link Through Time," *Cincinnati Enquirer*, April 27, 1992. For further technical information on the bridge see *Historic Bridges Conference, October 23 and 24, 1997, Cincinnati Museum Center, Cincinnati, Ohio* (Columbus, Ohio: Burgess & Niple, Ltd., 1997), esp. William E. Worthington, Jr., "John A. Roebling and the Cincinnati Bridge," 12–21, and Joseph F. Gastright, "Wilhelm Hildenbrand and the 1895 Recon-

struction of the Roebling Suspension Bridge," 24–35.

32. Cited in Mike Rutledge, "$6 Million Paint Job for Bridge," *Cincinnati Post*, May 18, 2004. This article dealt with the need for repainting of the bridge and what the estimated costs would be.

33. "Roebling Ready to Reopen," *Cincinnati Post*, March 22, 2007. Cost for painting were projected at $6.5 million.

34. See the editorial: "Kentucky Keeps a Rusting Roebling on Hold," *Cincinnati Enquirer*, June 22, 2004.

FOR **FURTHER** READING

Most works about Roebling concentrate on the Brooklyn Bridge, such as D. B. Steinman, *The Builders of the Bridge: The Story of John Roebling and His Son* (New York: Harcourt, Brace and Company, 1945); and David McCullough, *The Great Bridge* (New York: Simon & Schuster, 1972). There is only one book dealing specifically with the Ohio Bridge, Harry R. Stevens, *The Ohio Bridge* (Cincinnati: Ruter Press, 1939), but it concentrates on bridge construction and the bridge company, rather than Roebling and his legacy.

For biographical information about Roebling, see Hamilton Schuyler, *The Roeblings: A Century of Engineers, Bridgebuilders and Industrialists* (Princeton, N.J.: Princeton University Press, 1931); Gustav Koerner, *Das deutsche Element in den Vereinigten Staaten von Nordamerika, 1818–1848* (Cincinnati: A. E. Wilde, 1880); and H. J. Ruetenik, *Berühmte deutsche Vorkämpfer für Fortschritt, Freiheit und Friede in Nord-Amerika* (Cleveland, Ohio: Forest City Bookbinding Co., 1899).

Other general works dealing with the region are John Clubbe, *Cincinnati Observed: Architecture and History* (Columbus: Ohio State University Press, 1992); B. Drake and E. D. Mansfield, *Cincinnati in 1826* (Cincinnati: Printed by Morgan, Lodge, & Fisher, 1827); Barbara Keyser Gargiulo, ed. *Hamilton*

County, Ohio: As Extracted From Henry Howe's Historical Collections of Ohio (Milford, Ohio: Little Miami Publishing Co., 2005); Geoffrey J. Giglierano and Deborah A. Overmyer, *The Bicentennial Guide to Greater Cincinnati: A Portrait of Two Hundred Years* (Cincinnati: Cincinnati Historical Society, 1988); John Kleber, ed., *The Kentucky Encyclopedia* (Lexington, Ky.: University Press of Kentucky, 1992); *The Fifth Historic Bridges Conference* (Columbus, Ohio: Burgess & Niple, Ltd., 1997); *The WPA Guide to Cincinnati, With a New Introduction by Zane L. Miller and a New Preface by Harry Graff* (Cincinnati: Cincinnati Historical Society, 1987); and Don Heinrich Tolzmann, *German Heritage Guide to the Greater Cincinnati Area*, 2nd Ed. (Milford, Ohio: Little Miami Publishing Co., 2007).

INDEX

ABOUT THE **AUTHOR**

Dr. Tolzmann is shown standing next to the Cincinnati monument that honors Friedrich Hecker, a leader of the 1848 Revolution in Germany. In Cincinnati, Hecker and other Forty-eighters established the first Turner Society in America.

DON HEINRICH TOLZMANN is the Curator of the German-Americana Collection and Director of the German-American Studies Program at the University of Cincinnati. He is the author and editor of numerous works relating to German-American history and culture, including several for the Cincinnati and Northern Kentucky area.

OTHER **TITLES** BY THIS AUTHOR
published by LIttle Miami Publishing include

German Pioneer Accounts of the Great Sioux Uprising of 1862, edited by Don Heinrich Tolzmann (2002)—The fascinating firsthand reports of two German-American women who were kidnapped as children and survived to tell the story.

New Ulm, Minnesota: J. H. Strasser's History and Chronology, translated and edited by Don Heinrich Tolzmann (2003)—The history of a frontier settlement founded by Cincinnati Germans in the 1850s.

Wooden Shoe Hollow: Charlotte Pieper's Cincinnati German Novel, edited by Don Heinrich Tolzmann (2004)—The classic German-American historical novel about the community known as Wooden Shoe Hollow.

Missouri's German Heritage (2004) and *Missouri's German Heritage*, 2nd Edition, edited by Don Heinrich Tolzmann (2006)—Explore the early German history of Missouri through the translated and edited writings of Gustav Koerner and learn about the German pioneers, Gottfried Duden and Friedrich Muench, by Dorris Keeven Franke and Siegmar Muehl.

German Heritage Guide to the State of Ohio, by Don Heinrich Tolzmann (2005)—A timeline of events relating to the German immigration and settlement in Ohio, influential people and their contributions, current sites open to visitors for tours or reenactments, and a list of libraries and museums in the state that provide additional resources for research.

Illinois' German Heritage, edited by Don Heinrich Tolzmann (2005)—Explores the rich German heritage of Illinois from the early 19th century to the present in this first historical survey of the state's German element.

German Heritage Guide to the Greater Cincinnati Area, 2nd Edition, by Don Heinrich Tolzmann (2007)—A guide that explores the rich German Heritage of the Cincinnati; Hamilton County, Ohio; Butler County, Ohio; Northern Kentucky; and Indiana regions, among others. This new edition includes a resource chapter for radio programs, restaurants, bakeries, churches, and many other places of interest.